THE COMMONWEALTH AND INTERNATIONAL LIBRARY
Joint Chairmen of the Honorary Editorial Advisory Board
SIR ROBERT ROBINSON, O.M., F.R.S., LONDON
DEAN ATHELSTAN SPILHAUS, MINNESOTA
Publisher: ROBERT MAXWELL, M.C., M.P.

STRUCTURES AND SOLID BODY MECHANICS DIVISION
General Editor: B.G. NEAL

ANALYSIS AND DESIGN
OF STRUCTURAL SANDWICH PANELS

ANALYSIS AND DESIGN OF STRUCTURAL SANDWICH PANELS

by
HOWARD G. ALLEN

Senior Lecturer, Dept. of Civil Engineering, The University, Southampton

PERGAMON PRESS

OXFORD · LONDON · EDINBURGH · NEW YORK
TORONTO · SYDNEY · PARIS · BRAUNSCHWEIG

Pergamon Press Ltd., Headington Hill Hall, Oxford
4 & 5 Fitzroy Square, London W.1
Pergamon Press (Scotland) Ltd., 2 & 3 Teviot Place, Edinburgh 1
Pergamon Press Inc., Maxwell House, Fairview Park, Elmsford,
New York 10523
Pergamon of Canada Ltd., 207 Queen's Quay West, Toronto 1
Pergamon Press (Aust.) Pty. Ltd., 19a Boundary Street,
Rushcutters Bay, N.S.W. 2011, Australia
Pergamon Press S.A.R.L., 24 rue des Écoles, Paris 5^e
Vieweg & Sohn GmbH, Burgplatz 1, Braunschweig

Copyright © 1969 Howard G. Allen

All Rights Reserved. No part of this publication may be reproduced, stored in a retrieval system, or transmitted, in any form or by any means, electronic, mechanical, photocopying, recording or otherwise, without the prior permission of Pergamon Press. Ltd.,

First edition 1969

Reprinted 1969

Library of Congress Catalog Card No. 68-24060

Printed in Hungary
and reprinted lithographically
by Compton Printing Ltd., Aylesbury & London

This book is sold subject to the condition
that it shall not, by way of trade, be lent,
resold, hired out, or otherwise disposed
of without the publisher's consent,
in any form of binding or cover
other than that in which
it is published

08 012869 8 (flexicover)
08 012870 X (hard cover)

CONTENTS

PREFACE ... IX

LIST OF PRINCIPAL SYMBOLS AND DEFINITIONS ... XII

1. Introduction ... 1

2. Sandwich Beams ... 8

    * 2.1. Sandwich Beams: Application of Ordinary Beam Theory ... 8
    * 2.2. Sign Convention for Bending of Beams ... 14
    * 2.3. Deflection of a Simply-supported Sandwich Beam with Antiplane Core and Thin Faces (Symmetrical Load) ... 15
    2.4. Deflection of a Simply-supported Sandwich Beam with Antiplane Core and Thin Faces (Unsymmetrical Load) ... 19
    2.5. Deflections and Stresses in a Sandwich Beam with Antiplane Core and Thick Faces ... 21
    2.6. Simply-supported Beam with Central Point Load W (Antiplane Core and Thick Faces) ... 25
    2.7. Simply-supported Beam with Uniformly Distributed Load (Antiplane Core and Thick Faces) ... 33
    2.8. Beam with Four-point Loading (Antiplane Core and Thick Faces) ... 36
    2.9. Sandwich Beams with Faces of Unequal Thickness (Antiplane Core) ... 41
    2.10. Sandwich Beams in which the Modulus of Elasticity of the Core Parallel with the Axis is not Small (Faces of Equal Thickness) ... 43
    * 2.11. Wide and Narrow Beams ... 46

---

* Sections marked with an asterisk are of an elementary nature and may generally be read independently of the other parts of the book.

3. Buckling of Sandwich Struts 48

* 3.1. Buckling of Pin-ended Sandwich Strut with Antiplane Core and Thin Faces 48
* 3.2. Buckling of Pin-ended Sandwich Strut with Antiplane Core and Thick Faces 50
* 3.3. Further Consideration of Buckling (Antiplane Core, Thick Faces) 53
* 3.4. Wrinkling Instability 56

4. Analysis of Sandwich Beams and Struts by Strain Energy Methods 57

* 4.1. Introduction, Notation, Assumptions 57
* 4.2. Displacements and Strains 59
* 4.3. Strain Energy 61
* 4.4. Simply-supported Beam-column with Sinusoidal Transverse Load 65
* 4.5. Evaluation of Stresses due to Combined End-load and Sinusoidal Transverse Load 68
* 4.6. Deflections and Stresses due to Non-sinusoidal Transverse Loads 71

5. Bending and Buckling of Isotropic Sandwich Panels with Very Thin Identical Faces (Ritz Method) 76

* 5.1. Introduction 76
* 5.2. Displacements and Strains 78
* 5.3. Strain Energy 80
* 5.4. Plate with Isotropic Faces and Core 83
* 5.5. Other Types of Simply-supported Sandwich Plate 95
* 5.6. Boundary Conditions for a Simply-supported Panel 97

6. Bending and Buckling of Orthotropic Sandwich Panels with Thick Dissimilar Faces (Ritz Method) 99

6.1. Introduction 99
6.2. Displacements and Strains 100
6.3. Strain Energy 104
6.4. Potential Energy of Applied Forces 107
6.5. Minimization of Total Energy 107
6.6. Procedure in Particular Cases 114
6.7. Simply-supported Sandwich Plate with Identical Isotropic Thick Faces and "Isotropic" Core 116

7. Bending and Buckling of Orthotropic Sandwich Panels with Thin Faces; Alternative Solution Based on Differential Equations of Sandwich Panel 126

7.1.	Introduction	126
7.2.	Notation, Assumptions and Basic Equations	127
7.3.	General Differential Equations	131
7.4.	Simplified Form of Differential Equations	133
7.5.	Core Shear Strains and the Effect of Face Thickness	135
7.6.	Evaluation of the Stiffnesses D_x, D_{Qx}, etc.	137
7.7.	Solution for Simply-supported Orthotropic Panel with Edge Loads and Sinusoidal Transverse Load	139
7.8.	Buckling of Simply-supported Orthotropic Panel	143
7.9.	Solution for Simply-supported Orthotropic Panel with Edge Loads and Uniform Transverse Load	147

8. Wrinkling and Other Forms of Local Instability 156

* 8.1.	Introduction	156
* 8.2.	Long Strut Supported by a Continuous Elastic Isotropic Medium	157
* 8.3.	Wrinkling of Faces of Sandwiches with Isotropic Cores of Finite Thickness	160
8.4.	Behaviour of the Core during Wrinkling	167
8.5.	The Winkler Hypothesis	169
8.6.	Stability of Faces Attached to Antiplane Core with Infinite Stiffness Perpendicular to the Faces	171
8.7.	Initial Irregularities of the Faces	173
* 8.8.	Some Special Observations on Wrinkling Behaviour	178
* 8.9.	Local Instability of the Elements of a Sandwich, other than Wrinkling	179
8.10.	Interaction of Wrinkling and Overall Instability	181

9. The Development of the Theory of Sandwich Panels 190

9.1.	Historical Development of Sandwich Theory	190
9.2.	Comparison of Some Common Notations	200
9.3.	Boundary Conditions	207
9.4.	Panels with Edges which are not Simply-supported	208
9.5.	Shear and Other Edge-wise Loads	211
9.6.	Large Deflections	212
9.7.	Initial Deformations	213

10. Formulae for Analysis 217

*10.1. *Assumptions and Approximations* 217
*10.2. *Sandwich Beams* 220
*10.3. *Sandwich Struts* 221
*10.4. *Wrinkling* 223
*10.5. *Buckling of Simply-supported Panels with Edge Forces in the x-direction* 226
*10.6. *Bending of Simply-supported Panels with Uniform Transverse Load* 229
*10.7. *Bending of Simply-supported Panels with Uniform Transverse Load Combined with Edge Forces in the x-direction* 232
*10.8. *Modifications for Faces of Unequal Thickness or Dissimilar Materials* 232
*10.9. *Buckling and Bending of Panels with Other Types of Load or Boundary Conditions* 233

11. Design of Sandwich Beams, Struts and Panels 235

*11.1. *Introduction* 235
*11.2. *Determination of Core Thickness* 237
*11.3. *Optimum Design: Determination of Core and Face Thickness for Minimum Weight (or Cost)* 241

12. Properties of Materials Used in Sandwich Construction: Methods of Testing 245

*12.1. *Face Materials* 245
*12.2. *Core Materials and Their Properties* 248
*12.3. *Methods of Testing Core Materials* 254
*12.4. *Tests on Sandwich Constructions* 257

* Appendix I. Properties of Isotropic and Orthotropic Elastic Solids 264

Appendix II. Differential Equation for a Sandwich Beam-column 268

REFERENCES 271

INDEX 281

PREFACE

THIS book is intended to appeal to those engineers who are faced for the first time with the arduous task of finding a way through the tangled literature on sandwich theory.

It is not, however, an encyclopaedia of sandwich construction. The aim has been not to mention every paper on the topic which has appeared in print, but to provide a simple guide to the principal aspects of the theory of sandwich construction and to the assumptions on which it is based. This has entailed a very selective interpretation of the literature and a number of advanced topics have been omitted in order to do justice to the essentials.

The opening chapters deal with the relatively simple problems of sandwich beams and struts; these chapters (and especially Chapter 4) serve as an introduction to the later ones on panels. The treatment throughout is in engineering terms. For example, the discussion of the bending of sandwich beams in Chapter 2 grows naturally from the ordinary theory of bending; it does not start (as it might) with a general consideration of the equilibrium of elements of the core and the faces and with the compatibility of core and face deformations. Mathematical rigour has been sacrificed to a clear interpretation of the physical realities of sandwich behaviour. For instance, emphasis is given to the real significance of the assumptions made in sandwich theory concerning the thicknesses and relative stiffnesses of the faces and the core.

The bending and buckling of sandwich panels is introduced in Chapter 5. It is hoped that the simplified presentation will make the material intelligible even to readers who hitherto have had little acquaintance with the theory of bending of plates and with the use of strain energy for such problems. The more thorough panel analysis in Chapter 6 is merely an extension of the preliminary work of Chapter 5.

The panel analyses in Chapters 5 and 6 (based on the Ritz method) and in Chapter 7 (based on the derivation of differential equations for a sandwich plate) are in many ways equivalent and the designer may adopt the technique which pleases him best. The equivalence of the two methods, and of others, is demonstrated in Chapter 9.

Within the limits set by its introductory nature, the book has been arranged to provide information of immediate practical value. This is especially true of Chapter 10, in which are summarized the results derived in other chapters and which may be used without detailed reference to the rest of the book. Furthermore, the contents list contains an indication of those parts of the book which are of an introductory or elementary character and which are suitable for a first reading.

Most of the literature on sandwich construction is concerned with aerospace applications but there is also a considerable interest among the building industry in the possibilities of structural and semi-structural panels. Unlike those for aircraft, panels designed for use in buildings and for other semi-structural purposes often tend to be designed with rather thick faces and weak cores. In such circumstances the assumption that the faces are thin (often made in aircraft papers) may not be valid. The validity of the assumption and the consequences of avoiding it are explored at greater length than usual. For this reason the book should be of interest not only to aeronautical engineers but also to anyone concerned with the design of sandwich panels in the building, plastics and boat-building industries.

The information on which this book is based is taken almost entirely from readily available British or American publications. This is purely a matter of convenience and it should be noted that almost parallel information exists in other languages, particularly Russian and Dutch. Although great initial impetus was given to the development of sandwich theory by the wartime investigations of British workers, anyone who writes a book on sandwich construction must find it impossible to avoid repeated reference to the very extensive post-war studies of sandwich construction performed at the U.S. Forest Products Laboratory, which must surely come first in any list of acknowledgements. In particular I am grateful to FPL for permission to use fig. 16 of Report 1810 as the basis of my Fig. 8.11.

The many plate-buckling graphs which have appeared in FPL reports and elsewhere have served as models for the equivalent graphs in this book. Nevertheless, all of my own graphs for the bending and buckling of sandwich plates are based on original calculations. I owe a great debt to Mr. G. M. Folie, who prepared the computer programmes for this work.

Finally I wish to express my thanks to the Royal Aeronautical Society (for permission to use figs. 5, 13 and 14 of ref. 10.1 as the basis of Figs. 8.5 and 8.6), to the American Institute of Aeronautics and Astronautics (for permission to use fig. 4 of ref. 9.3 in the preparation of Figs. 8.13 and 8.14) and to the McGraw-Hill Book Company (for permission to copy fig. 2.10 of ref. 35.15 in the outlines of Fig. 9.4).

LIST OF PRINCIPAL SYMBOLS AND DEFINITIONS

Elasticity

x, y, z	Rectangular coordinates. z is normal to the faces, positive downwards. In a beam or strut, x lies along the axis of the member.
u, v, w	Displacements in the x-, y-, z-directions.
e_x, e_y, e_z	Tensile strains ⎫
$\sigma_x, \sigma_y, \sigma_z$	Tensile stresses ⎬ In x-, y-, z-directions.
E_x, E_y, E_z	Moduli of elasticity[†] ⎭
$\gamma_{zx}, \gamma_{xy}, \gamma_{yz}$	Shear strains ⎫ In zx-, xy-,
$\tau_{zx}, \tau_{xy}, \tau_{yz}$	Shear stresses ⎬ yz-planes.
G_{zx}, G_{xy}, G_{yz}	Moduli of rigidity (shear moduli)[†] ⎭
ν_x, ν_y	Poisson's ratios in the xy-plane. When σ_x is the only non-zero stress, $\nu_x = -e_y/e_x$.
$g = 1 - \nu_x \nu_y$	

Beams and Struts

b	Width of beam	⎫
t	Face thickness	⎪
d	Distance between centre-lines of opposite faces $= c + t$	⎬ Figs. 2.1, 2.17.
h	Overall thickness of sandwich	⎪
L	Span	⎭
I	Second moment of area, especially total value of faces about centroid of sandwich beam	Eqn. (2.19).

[†] Suffixes are omitted where the material is isotropic and where there is no ambiguity. It is usually obvious which symbols refer to the properties of the faces and which to the properties of the core. Where it is not, the suffixes f and c are used.

LIST OF PRINCIPAL SYMBOLS AND DEFINITIONS xiii

I_f Sum of second moments of area of faces about their own separate centroidal axes } Eqn. (2.19).

q Distributed transverse load on beam, per unit length
Q Shear force
M Bending moment } Fig. 2.2.

Δ Displacement at a particular point.
W Point load.
D Flexural rigidity of beam or strut (see Table on p. xv).
P Axial load in strut.

MISCELLANEOUS

Suffixes 1 and 2   (i) Upper and lower faces respectively.
                             (ii) Primary and secondary transverse displacements (Section 2.5).

$A = b\,d^2/c$

PANELS

a, b Length and width of panel, in x and y directions.

$M_x, M_y, M_{xy}, (M_{yx})$ Bending and twisting moments
$N_x, N_y, N_{xy}, (N_{yx})$ Tensile and shearing membrane forces
Q_x, Q_y Shear forces } Per unit width Fig. 5.2.

q Transverse load per unit area.

D_x, D_y, D_{xy} Bending and twisting stiffnesses
D_{Qx}, D_{Qy} Shear stiffnesses } Per unit width Definitions, Section 7.2.

B_{Qx}, B_{Qy} Modified shear stiffnesses
$(B_{Qx} = b^2 D_{Qx};\ B_{Qy} = b^2 D_{Qy})$ } Formulae, Section 7.6.

xiv LIST OF PRINCIPAL SYMBOLS AND DEFINITIONS

v_x, v_y "Poisson's ratios" defined in terms of curvature. Equations (7.1), (7.2).

$g = 1 - v_x v_y$

m, n Positive integer suffixes denoting mode of deformation, usually with m half-waves in x-direction, n half-waves in y-direction.

λ If a line is drawn through the core of an undeformed sandwich, normal to the middle plane, then, as the sandwich deforms, the line rotates in the zx-plane through an angle $\lambda(\partial w/\partial x)$ (Fig. 4.1).

μ Similarly, $\mu(\partial w/\partial y)$ is the rotation in the yz-plane.

P Compressive edge load per unit width.

K Non-dimensional plate buckling coefficient. Suffixes 1, 2 and 3 merely distinguish the values used in the comparable equations (5.29), (6.36) and (7.57).

WRINKLING

l Half-wavelength.

B_1 Non-dimensional buckling coefficient in the basic wrinkling formula, $\sigma = B_1 E_f^{\frac{1}{3}} E_c^{\frac{2}{3}}$ (equation (8.12)).

B_1' Non-dimensional buckling coefficient in the alternative wrinkling formula, $\sigma = B_1'(E_f E_c G_c)^{\frac{1}{3}}$ (equation (10.16)).

$$B_1' = B_1[2(1+v_c)]^{\frac{1}{3}}.$$

B_2 Non-dimensional coefficient in formula for face failure associated with initial irregularities,

$$\sigma = B_2 E_f^{\frac{1}{3}} E_c^{\frac{2}{3}} \quad \text{(equation (8.34))}.$$

ϱ Non-dimensional coefficient which defines character of sandwich for purpose of wrinkling analysis,

$$\varrho = \frac{t}{c}\left\{\frac{E_f}{E_c}\right\}^{\frac{1}{3}}$$

Isotropic Panels, Beams, Struts

	Beam or strut	Panel
Flexural rigidity	D	D
Flexural rigidity, neglecting local bending stiffness of faces	$D_1 = E_f b t d^2 / 2$	$D_2 = E_f t d^2 / 2(1 - v_f^2)$ (cylindrical bending) $D_x = D_y = E_f t\, d^2/2$ (anticlastic bending)
	n^2	$\Omega = \dfrac{m^2 b^2}{a^2} + n^2$
Nominal ratio of flexural rigidity to core shear stiffness	$\xi = \dfrac{\pi^2}{2} \dfrac{E_f}{G_c} \dfrac{ct}{L^2}$ $= \dfrac{\pi^2}{L^2} \dfrac{E_f b t d^2}{2} \dfrac{c}{G_c b d^2}$	$\varrho = \dfrac{\pi^2}{2(1-v_f^2)} \dfrac{E_f}{G_c} \dfrac{ct}{b^2}$ $= \dfrac{\pi^2}{b^2} \dfrac{E_f t d^2}{2(1-v_f^2)} \dfrac{c}{G_c d^2}$
Special case of above when faces are very thin; $c = d$. See Chapter 5.	$\xi = \dfrac{\pi^2}{2} \dfrac{E_f}{G_c} \dfrac{dt}{L^2}$	$\varrho = \dfrac{\pi^2}{2(1-v_f^2)} \dfrac{E_f}{G_c} \dfrac{dt}{b^2}$
Factor for sandwiches with thick faces, Chapters 4 and 6	$\chi = \left[\dfrac{1}{1 + \xi n^2} + \dfrac{t^2}{3d^2} \right]^{-1}$	$\zeta = \left[\dfrac{1}{1 + \varrho \Omega} + \dfrac{t^2}{3d^2} \right]^{-1}$

Definitions

Antiplane core One in which $\sigma_x = \sigma_y = \tau_{xy} = 0$. As a consequence, the shear stresses τ_{zx}, τ_{yz} are independent of z.

Core shear stiffness $AG = G b d^2 / c$ (for beam of width b).

Very thin face One with negligible stiffness in bending about its own axis, and sufficiently thin for d to be equated with c.

Thin face	One with negligible stiffness in bending about its own axis, but not so thin that d and c may be equated.
Thick face	One with significant stiffness in bending about its own axis, and too thick for d and c to be equated. See Section 10.1 for quantitative assessments of face thicknesses.

CHAPTER 1

INTRODUCTION

ALTHOUGH the Second World War "Mosquito" aircraft is often quoted as the first major structure to incorporate sandwich panels, sandwich construction has been used in many earlier but less spectacular circumstances. Reviewers of the history of sandwich construction compete to name the first person to describe the principle and the record appears to be held by Fairbairn[35.1] (1849). It seems likely, however, that the idea of sandwich construction has occurred independently to many engineers at different times; no doubt a diligent researcher will eventually find it in the works of the ubiquitous Leonardo da Vinci.

The simplest type of sandwich consists of two thin, stiff, strong sheets of dense material separated by a thick layer of low density material which may be much less stiff and strong (Fig. 1.1a). As a crude guide to the proportions, an efficient sandwich is obtained when the weight of the core is roughly equal to the combined weight of the faces. Obviously the bending stiffness of this arrangement is very much greater than that of a single solid plate of the same total weight made of the same material as the faces.

The core has several vital functions. It must be stiff enough in the direction perpendicular to the faces to ensure that they remain the correct distance apart. It must be stiff enough in shear to ensure that when the panel is bent the faces do not slide over each other. If this last condition is not fulfilled the faces merely behave as two independent beams or panels and the sandwich effect is lost. The core must also be stiff enough to keep the faces nearly flat; otherwise it is possible for a face to buckle locally (wrinkle) under the influence of compressive stress in its own plane. The core *must* satisfy all these requirements and it is also important that the adhesive should not be sufficiently flexible to permit substantial relative movements of the faces and the core.

If the core is stiff enough it may make a useful contribution to the bending stiffness of the panel as a whole. This contribution

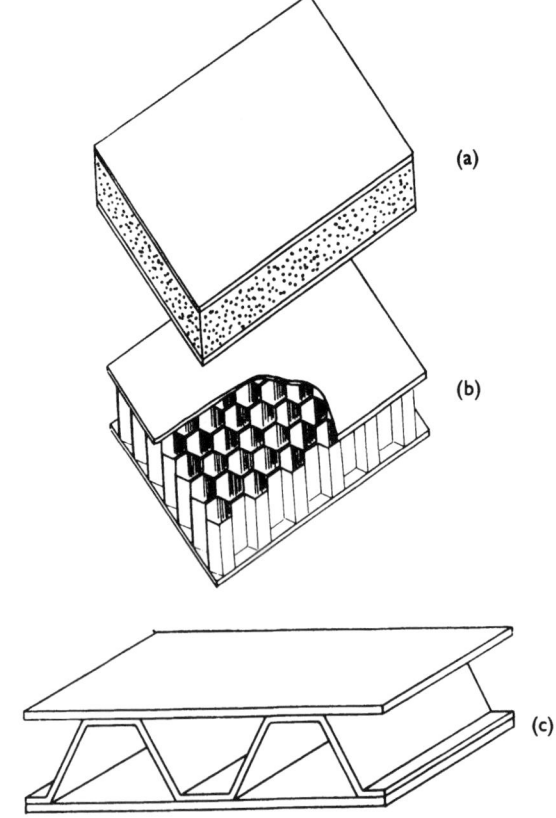

FIG. 1.1. Sandwich panels with (a) expanded plastic core, (b) honeycomb core, (c) corrugated core.

is rather small in the case of the low-density cores which are usually employed and it is very often expedient to ignore it. This also leads to a considerable simplification of the analysis of stresses and deflections.

Sandwich panels for aircraft structures almost invariably employ metal faces with metal honeycomb or corrugated cores. The honeycomb is formed from strips of thin aluminium alloy or steel foil deformed and joined together (Fig. 1.1b). The corrugated core is a fluted metal sheet attached alternately to the upper and lower faces (Fig. 1.1c). Sandwich construction is favoured for all but the heaviest load intensities because, unlike skin-stringer construction, it is relatively free from buckling deformations at working loads.

Panels for radomes, which must be permeable to radar waves, utilize glass-reinforced plastics for the faces and either the same material or resin-impregnated paper for the honeycomb core.

Panels for use in the building industry have hitherto been of a mainly semi-structural character, called upon to carry relatively small loads over fairly long spans. Building panels, like aircraft panels, should be light in weight but, unlike the aircraft panels, they must be cheap. All-metal panels may yet find substantial application in buildings but there is also great scope for many other materials. For faces there are asbestos cement, glass-reinforced plastics, plywood, glass-reinforced cement, plasterboard, resin-impregnated paper, hardboard and ferro-cement (as distinct from ordinary reinforced concrete). Non-metallic honeycomb and corrugated cores are made of resin-impregnated paper and reinforced plastics; "solid" cores are made of perforated chipboard, balsa wood, several kinds of expanded plastics, foamed glass, lightweight concrete and clay products. New materials and new combinations of old materials are constantly being proposed and used.

Certain variations on the idea of a honeycomb core are illustrated in Fig. 1.2. The interlocking straight strips in Fig. 1.2a

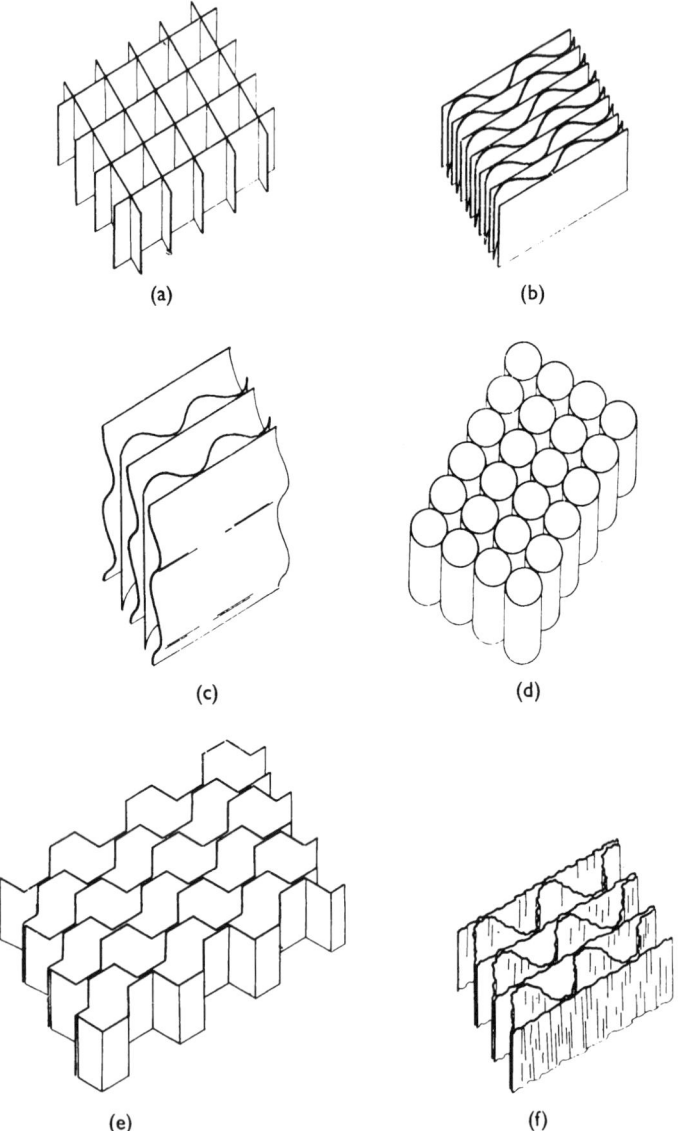

FIG. 1.2. Types of "honeycomb" core.

need no comment. The cores in Figs. 1.2b and 1.2c are made by gluing large numbers of flat and corrugated sheets together in different ways; the use of resin-impregnated paper in this way is described in refs. 31.1 and 31.2. Figure 1.2d was an early proposal for a substitute for balsa wood with the grain perpendicular to the faces.[13.4] Figure 1.2e is a deformed honeycomb, usually in aluminium alloy foil, which can easily be bent to fit the surface of a cylindrical mould.[11.9] Ordinary honeycomb tends to bend anticlastically and will not readily fit a cylindrical surface. The "multiwave" honeycomb[11.6] in Fig. 1.2f is fabricated from crinkled foil and it has the same useful characteristics as the deformed honeycomb.

Variations of the corrugated core are shown in Fig. 1.3. The simple parallel-strip arrangement of Fig. 1.3a is sometimes stiffened by the addition of expanded plastics to fill the voids. The tubular core of Fig. 1.3b and the double truss-core of Fig. 1.3c are rather rare. The dimpled core shown in Fig. 1.3d is similar in appearance to the pulpboard commonly used for packing eggs. Unlike the corrugated core it has similar properties in the two principal directions, but it retains the advantages of easy fabrication and good adhesion to the faces (by rivets or welding if desired).

The correct design of the details of sandwich construction is at least as important as the analysis of deflections, stresses and buckling loads. These details include nature of the edge members, splices and joints in the cores and faces, stiffeners and inserts to distribute concentrated loads, type of adhesive, method of fabrication and so forth. If the temperatures of the two faces differ, or if the moisture contents differ (as they may in asbestos cement or hardboard, for example) the differential expansion of the faces may lead to substantial transverse deflections. In building panels, especially, problems arise concerning acoustic insulation, vapour transmission and fire resistance (but not usually heat insulation). All of the factors mentioned can be very important design consid-

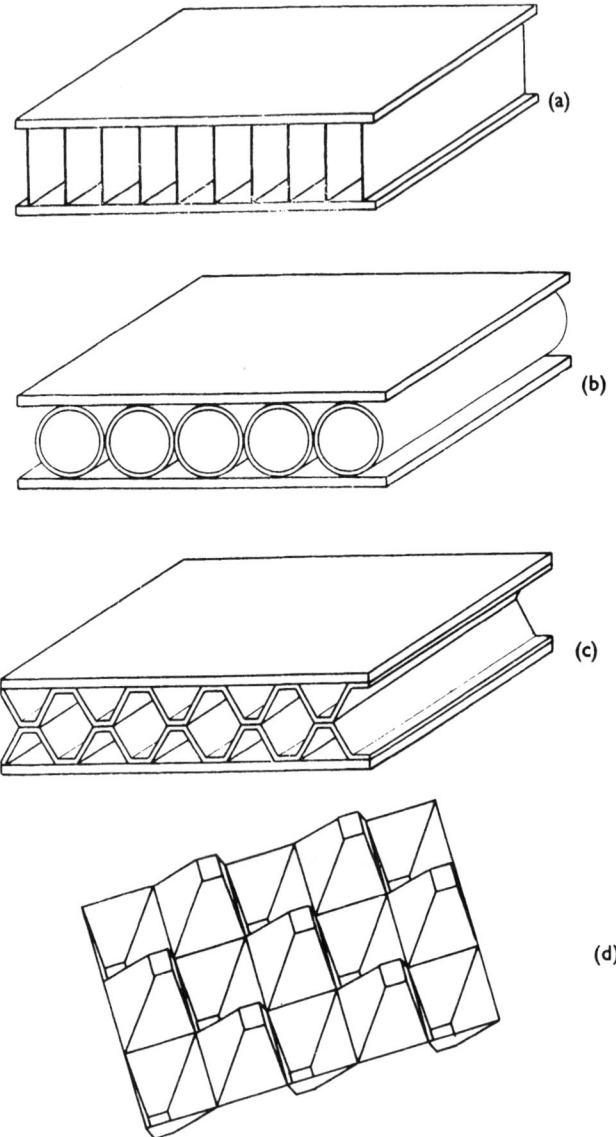

FIG. 1.3. Variations of the "corrugated" core.

erations but they are beyond the scope of a book of this kind. Nevertheless, much useful practical information can be found in refs. 5.12, 30.2, 31.1, 31.2, 35.9 and 35.12 and in other papers mentioned in the list of references at the end of this book. This list also includes a selection of reviews and bibliographies which may be used, if desired, to obtain information about many other papers not in the list.

CHAPTER 2

SANDWICH BEAMS

2.1. Sandwich Beams: Application of Ordinary Beam Theory

The sandwich beam illustrated in Fig. 2.1 consists of two thin skins or faces each of thickness t, separated by a thick layer, or core, of low density material of thickness c. The overall depth of the beam is h and the width is b. All three layers are firmly

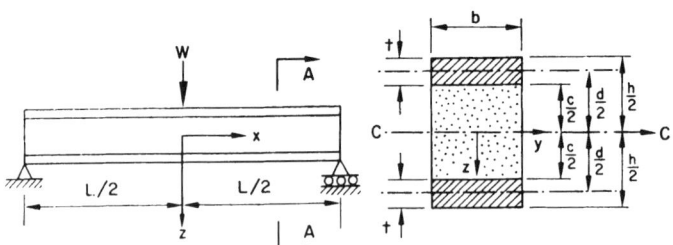

FIG. 2.1. Dimensions of sandwich beam. Section AA on right.

bonded together and the face material is much stiffer than the core material. It is assumed for the present that the face and core materials are both isotropic.

The stresses and deflections in a beam of this kind may be found, to a first approximation, by the use of the ordinary theory of bending. The theory is based on the assumption that cross-sections which are plane and perpendicular to the longitudinal axis of the unloaded beam remain so when bending takes place. This assumption leads to the well-known relationship between bending moment (M) and curvature ($1/R$):

$$\frac{M}{EI} = -\frac{1}{R}. \tag{2.1}$$

The negative sign is introduced to comply with the sign convention illustrated in Fig. 2.2. EI is the flexural rigidity which, in an ordinary beam, would be the product of the modulus of elasticity E and the second moment of area, I. It will be convenient

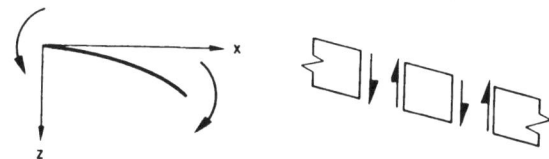

FIG. 2.2. Sign conventions. Left, positive deflection, slope and curvature; negative bending moment. Right, positive shear force, shear stress and shear strain.

to denote the flexural rigidity by the single symbol D. The sandwich beam in Fig. 2.1 is a composite beam, however, and its flexural rigidity is the sum of the flexural rigidities of the two separate parts, faces and core, measured about the centroidal axis cc of the entire cross-section. Thus:

$$D = E_f \cdot \frac{bt^3}{6} + E_f \cdot \frac{btd^2}{2} + E_c \cdot \frac{bc^3}{12}, \tag{2.2}$$

where E_f and E_c are the moduli of elasticity of the faces and core respectively and d is the distance between the centre lines of the upper and lower faces:

$$d = \frac{h+c}{2}. \tag{2.3}$$

(It is assumed for the present that the beam is narrow, so that stresses in the y-direction can be taken as zero.)

In the right-hand side of equation (2.2) the first two terms represent the stiffness of the faces associated with bending about the centroidal axis of the entire sandwich cc; of these, the first represents the local stiffness of the faces, bending separately about their own centroidal axes. The third term represents the bending stiffness of the core.

In practical sandwiches the second term is invariably dominant. The first term amounts to less than 1% of the second when

$$3\left(\frac{d}{t}\right)^2 > 100. \tag{2.4}$$

The error introduced by neglecting the first term is therefore negligible provided $d/t > 5.77$; sandwiches with thin metal faces usually satisfy this condition, but others with thick low-strength faces such as asbestos cement may not.

The third term amounts to less than 1% of the second (and may consequently be neglected) when

$$6\frac{E_f}{E_c}\frac{t}{c}\left(\frac{d}{c}\right)^2 > 100. \tag{2.5}$$

In many practical sandwiches $d/c \doteqdot 1$ and t/c lies in the range 0·02 to 0·1; if condition (2.5) is to be satisfied in these cases, the limiting value of E_f/E_c is somewhere between 835 and 167. It is by no means certain that ratios as high as these will be reached in every case.

The stresses in the faces and core may be determined by the use of ordinary bending theory, adapted to the composite nature of the cross-section. Because sections remain plane and perpendicular to the longitudinal axis† the strain at a point distant z below the centroidal axis cc (Fig. 2.1) is Mz/D. This strain may be multiplied by the appropriate modulus of elasticity to give the bending stress at the level z. For instance, the stresses in the faces

† It will be shown later that this assumption must be modified in certain circumstances.

and the core are, respectively,

$$\sigma_f = \frac{Mz}{D} E_f \quad \left(\frac{c}{2} \leqslant z \leqslant \frac{h}{2}; \quad -\frac{h}{2} \leqslant z \leqslant -\frac{c}{2}\right), \quad (2.6\text{a})$$

$$\sigma_c = \frac{Mz}{D} E_c \quad \left(-\frac{c}{2} \leqslant z \leqslant \frac{c}{2}\right). \quad (2.6\text{b})$$

The maximum face and core stresses are obtained with z equal to $\pm h/2$ and $\pm c/2$, respectively:

$$(\sigma_f)_{\max} = \pm \frac{ME_f}{D} \cdot \frac{h}{2} \qquad (\sigma_c)_{\max} = \pm \frac{ME_c}{D} \cdot \frac{c}{2}. \quad (2.7)$$

The ratio of the maximum face stress to the maximum core stress is $(E_f/E_c)(h/c)$. If the ultimate strengths of the face and core materials are exactly in proportion to their moduli the faces will fail marginally before the core does so, since h/c is slightly greater than unity.

The assumptions of the ordinary theory of bending lead to the common expression (2.8) for the shear stress, τ, in a homogeneous beam at a depth z, below the centroid of the cross-section:

$$\tau = \frac{QS}{Ib}. \quad (2.8)$$

Here Q is the shear force at the section under consideration, I is the second moment of area of the entire section about the centroid, b is the width at the level z_1 and S is the first moment of area of that part of the section for which $z > z_1$. The familiar distribution of such a shear stress in an I-beam is illustrated in Fig. 2.3.

For a compound beam such as the sandwich in Fig. 2.1 equation (2.8) must be modified to take account of the moduli of elasticity of the different elements of the cross-section:

$$\tau = \frac{Q}{Db} \sum (SE). \quad (2.9)$$

In this expression D is the flexural rigidity of the entire section and $\sum(SE)$ represents the sum of the products of S and E of all

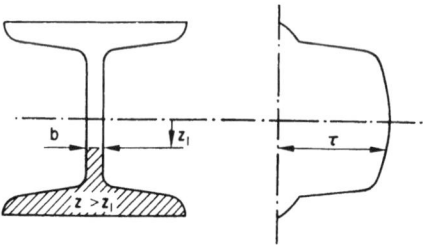

FIG. 2.3. Shear stress distribution in homogeneous *I*-beam.

parts of the section for which $z > z_1$. For example, if equation (2.9) is used to determine the shear stress at level z in the core of the sandwich in Fig. 2.1,

$$\sum(SE) = E_f \frac{btd}{2} + \frac{E_c b}{2}\left(\frac{c}{2} - z\right)\left(\frac{c}{2} + z\right).$$

The shear stress in the core is therefore

$$\tau = \frac{Q}{D}\left\{E_f \frac{td}{2} + \frac{E_c}{2}\left(\frac{c^2}{4} - z^2\right)\right\}. \tag{2.10}$$

An analogous expression may be obtained for the shear stress in the faces and the complete shear stress distribution across the depth of the sandwich is illustrated in Fig. 2.4a.

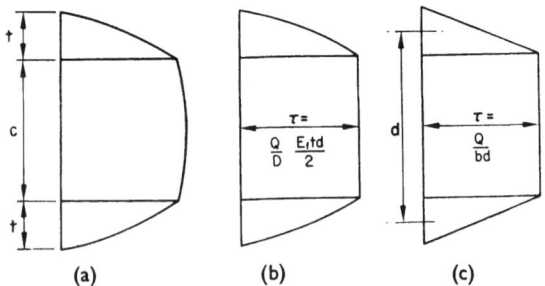

FIG. 2.4. Shear stress distribution in sandwich beam. (a) True shear stress distribution. (b) Effect of weak core (conditions (2.5) and (2.11) satisfied). (c) Effect of weak core, neglecting the local bending stiffnesses of the faces (conditions (2.4), (2.5) and (2.11) satisfied).

The ratio of the maximum core shear stress (at $z = 0$) to the minimum core shear stress (at $z = \pm c/2$) is

$$\left(1 + \frac{E_c}{E_f} \frac{1}{4} \frac{c^2}{td}\right).$$

This expression is within 1% of unity provided

$$4 \frac{E_f}{E_c} \frac{t}{c} \frac{d}{c} > 100. \tag{2.11}$$

Consequently, if condition (2.11) is satisfied, the shear stress may be assumed constant over the thickness of the core. Because d/c is usually near to unity, conditions (2.5) and (2.11) are roughly similar in effect. It may therefore be concluded that *where a core is too weak to provide a significant contribution to the flexural rigidity of the sandwich, the shear stress may be assumed constant over the depth of the core.*

For a weak core, therefore, it is permissible to write $E_c = 0$ in equations (2.2) and (2.10); the constant shear stress in the core is then given by:

$$\tau = \frac{Q}{D} \frac{E_f t d}{2}. \tag{2.12}$$

The way in which the shear stresses are distributed across the section is illustrated in Fig. 2.4b.

If, in addition, the flexural rigidity of the faces about their own separate axes is small (i.e. if condition (2.4) is fulfilled), then the first term on the right-hand side of equation (2.2) may be neglected as well as the third, leaving:

$$D = E_f \frac{btd^2}{2}. \tag{2.13a}$$

In this case equation (2.12) for the shear stress in the core reduces to the simplest possible form:

$$\tau = \frac{Q}{bd}. \tag{2.13b}$$

The corresponding shear stress distribution is illustrated in Fig. 2.4c. The difference between Figs. 2.4b and 2.4c is that in the latter the direct stress in each face is assumed to be uniform (because the local bending stress is neglected); it follows from this that the shear stress in the faces varies with depth in a linear fashion, not a parabolic one.

It is often convenient to invoke the concept of an "antiplane" core.† An antiplane core is an idealized core in which the modulus of elasticity in planes parallel with the faces is zero but the shear modulus in planes perpendicular to the faces is finite. (A honeycomb core is an approximation to an antiplane core.) By this definition $E_c = 0$ and the antiplane core makes no contribution to the bending stiffness of the beam. Conditions (2.5) and (2.11) are automatically satisfied and the shear stress distribution is similar to that shown in Fig. 2.4b.

Alternatively, if the equilibrium of a small element of an antiplane core is considered, it is easy to show that because no longitudinal stress acts upon it, the shear stress cannot vary with z.

2.2. Sign Convention for Bending of Beams

The sign conventions to be adopted for deflection, slope, curvature, bending moment and shear force are illustrated in Fig. 2.2. Loads and deflections (w) are measured positive downwards, in the direction of the z-axis. These particular sign conventions are compatible with the conventions adopted for panels in later chapters; the panel sign conventions are identical with those used in standard texts.[35.3, 35.14] As a result of the choice of sign convention it is necessary to introduce negative signs in some of the integral or differential relationships between distributed load (q), shear force (Q), bending moment (M), slope (dw/dx) and deflection (w). For reference, the full set of relationships, with the cor-

---

† Filon[35.2] used the term "antiplane" to describe a state of stress in which $\sigma_x = \sigma_y = \tau_{xy} = 0$.

rect signs, is given:

$$\left.\begin{array}{ll} \text{Deflection} & w \\ \text{Slope} & +\dfrac{dw}{dx} = w' \\ \text{Curvature} & +w'' \\ -M & +Dw'' \\ -Q & +Dw''' \\ +q & +Dw^{iv} \end{array}\right\} \quad (2.14)$$

2.3. Deflection of a Simply-supported Sandwich Beam with Antiplane Core and Thin Faces (Symmetrical Load)

Because the faces are thin, condition (2.4) is satisfied, the local bending stiffness of the faces is small and the first term on the right-hand side of equation (2.2) can be neglected. Because the core is antiplane, conditions (2.5) and (2.11) are satisfied and the third term on the right-hand side of equation (2.2) is negligible; also, the shear stress is constant throughout the depth of the core at any given section.

Consequently the flexural rigidity of the sandwich and the shear stress in the core are defined by equations (2.13); the shear stress distribution appears as in Fig. 2.4c.

In the first instance the transverse displacements of the beam (w_1) may be calculated by the ordinary theory of bending, using the relationships (2.14). For example, Fig. 2.5b shows the bending deformation of a simply supported beam with a central point load W. The points $a, b, c, \ldots$ lie on the centre-lines of the faces and the cross-sections $aa, bb, cc, \ldots$ rotate but nevertheless remain perpendicular to the longitudinal axis of the deflected beam. It is obvious that the upper face is compressed as the points $a, b, c, \ldots$ move closer together, while the lower face is placed in tension.

The shear stress in the core at any section is $\tau = Q/bd$ (equation (2.13b)). This is associated with a shear strain $\gamma = Q/Gbd$ which, like τ, is constant through the depth of the core; G is the modulus of rigidity of the core material. These shear strains lead to a new kind of deformation illustrated in Fig. 2.5c. The points $a, b, c, \ldots$,

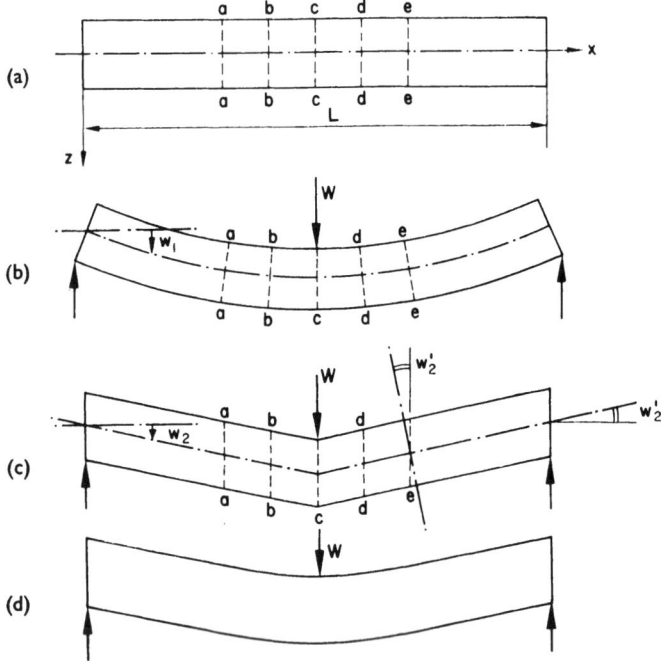

FIG. 2.5. Deflection of sandwich beam.

which lie on the centre-lines of the faces, do not move horizontally† but are displaced vertically by an amount w_2. The faces and the longitudinal centre-line of the beam tilt, however, and the relationship between the slope of the beam, dw_2/dx, and the core shear strain γ may be obtained from Fig. 2.6. In this diagram,

† Because the points $a, b, c, \ldots$ do not move horizontally, the new displacement does not alter the mean stress in the faces.

which shows an elevation of a short length of the sandwich, the distance de is equal† to $d(dw_2/dx)$. It is also equal to cf, which in turn is equal to γc.

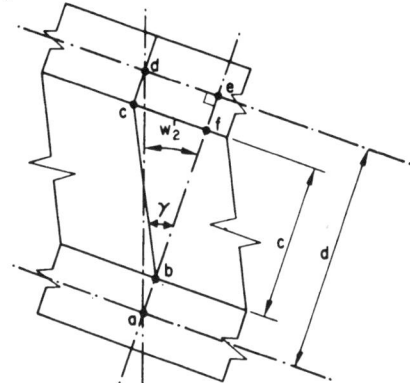

FIG. 2.6. Shear deformation of a beam with thick faces.

Hence
$$\frac{dw_2}{dx} = \gamma \frac{c}{d} = \frac{Q}{Gbd} \cdot \frac{c}{d} = \frac{Q}{AG}, \quad (2.15a)$$
where
$$A = bd^2/c. \quad (2.15b)$$

The product AG is often referred to as the shear stiffness of the sandwich. The displacement w_2, associated with shear deformation of the core, may be obtained by integration of equation (2.15a) in any particular problem.

For example, in the simply supported beam with a central point load W the shear force Q in the left-hand half of the beam is $+W/2$. Integration of equation (2.15a) with $Q = +W/2$ provides the displacement:

$$w_2 = \frac{W}{2AG} \cdot x + \text{constant} \qquad 0 \leq x \leq L/2.$$

† Shear strains in the faces are assumed to be negligible throughout the book.

The constant vanishes because $w_2 = 0$ at $x = 0$. The maximum value of w_2 occurs at the centre of the beam, $x = L/2$, and is equal to:

$$\varDelta_2 = \frac{WL}{4AG}.$$

The total central deflection $\varDelta$ is therefore the ordinary bending displacement $\varDelta_1$ with the displacement $\varDelta_2$ superimposed:

$$\varDelta = \varDelta_1 + \varDelta_2 = \frac{WL^3}{48D} + \frac{WL}{4AG}.$$

In general the displacement of any statically-determinate symmetrically-loaded sandwich beam with an antiplane core and thin faces may be found by similarly superimposing the "bending" and "shear" deflections w_1 and w_2. The bending deflections are found in the usual way and the shear deflections by integrating equation (2.15a).

It may be convenient to integrate equation (2.15a) in general terms with the following result:

$$w_2 = \frac{M}{AG} + \text{constant.} \tag{2.16}$$

For a simply supported beam with the origin at one support the constant is always zero. Consequently the shear displacement diagram is the same as the bending moment diagram, with a factor $1/AG$ applied to it.

For example, a simply supported beam of span L with a uniformly distributed load q has a central bending deflection $\varDelta_1$ equal to $+5qL^4/384D$, a standard result. The bending moment at the centre is $+qL^2/8$ and the central shear deflection $\varDelta_2$ is therefore $+qL^2/8AG$. The total deflection $\varDelta$ at the centre is given by

$$\varDelta = \varDelta_1 + \varDelta_2 = \frac{5qL^4}{384D} + \frac{qL^2}{8AG}.$$

2.4. Deflection of a Simply-supported Sandwich Beam with Antiplane Core and Thin Faces (Unsymmetrical Load)

In the previous section it was assumed that during shear deformation all points on the centre-lines of the faces moved only in the vertical direction, as in Fig. 2.5c. In general, however, it is possible for one face as a whole to move horizontally with respect to

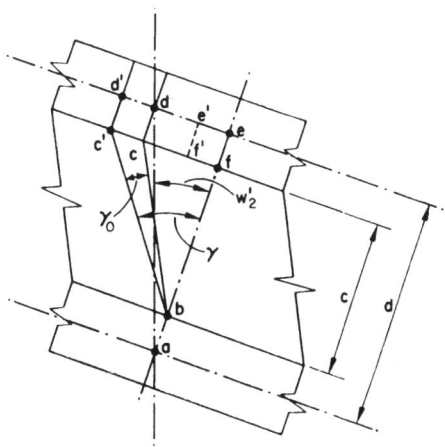

FIG. 2.7. Effect of γ_0 on shear deformation.

the other. The effect is illustrated in Fig. 2.7, which is similar to Fig. 2.6 in showing the axis of the beam at an angle w'_2 to the horizontal as a result of pure shear deformation of the core. However, the upper face has also been displaced to the left, so that the points $cdef$ in Figs. 2.6 and 2.7 now appear in new positions at $c'd'e'f'$. The angle cbc' is denoted by γ_0 and the following relationships exist:

$$cf = c'f - c'c = (\gamma - \gamma_0) \times c = de = w'_2 \times d.$$

Hence

$$w'_2 = (\gamma - \gamma_0)\frac{c}{d}. \qquad (2.17a)$$

Or

$$w'_2 = \frac{Q}{AG} - \gamma_0 \frac{c}{d}. \qquad (2.17b)$$

Or

$$w_2 = \frac{M}{AG} - \gamma_0 x \frac{c}{d} + \text{constant}. \qquad (2.17c)$$

Equations (2.15a) and (2.16) are merely special cases of (2.17b) and (2.17c).

Consider, for example, a simply-supported beam with a couple M_0 applied at one end (Fig. 2.8). The bending moment at x is $-M_0 x/L$, which value may be inserted in equation (2.17c):

$$w_2 = -\frac{M_0 x}{AGL} - \gamma_0 x \frac{c}{d} + \text{constant}.$$

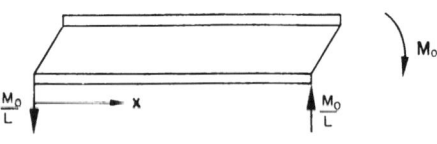

Fig. 2.8.

The boundary conditions $w_2 = 0$ at $x = 0, L$ show that the constant vanishes and γ_0 is equal to $-M_0 d/AGLc$. Substitution for γ_0 in equation (2.17c) shows that the transverse shear displacement w_2 is everywhere zero. However, all the sections through the core have rotated through an angle γ_0 as in Fig. 2.8. The shear strain γ at all points in the core is given by equation (2.17a) as

$$\gamma = \frac{d}{c} w'_2 + \gamma_0 = \gamma_0 = -\frac{M_0}{AGL} \cdot \frac{d}{c}.$$

The rotation γ_0 is always zero when the beam is loaded in a symmetrical manner, or when the relative horizontal displacement of the faces is prevented, for example at a clamped end.

2.5. Deflections and Stresses in a Sandwich Beam with Antiplane Core and Thick Faces

In this section it is assumed that conditions (2.5) and (2.11) are applicable; the core consequently makes no contribution to the flexural rigidity of the sandwich, D, and the shear stress is constant through the depth of the core. Condition (2.4) is *not* assumed to be valid; the faces have significant local bending stiffness and no longer behave as flexible membranes. The shear stress distribution is as in Fig. 2.4b.

The face of a sandwich may be said to undergo local bending when it bends about its own centroidal axis rather than about the centroidal axis of the complete sandwich. A face may also undergo purely extensional deformation when it is subjected to a uniform tensile or compressive stress. When the sandwich bends as a whole (as in Fig. 2.5b) the faces partake of both kinds of deformation. The contribution of the local bending stiffness of the faces to the bending stiffness of the entire sandwich is represented by the first term on the right-hand side of equation (2.2).

In addition to this, however, the local bending stiffness of the faces has an effect on the shear deformation of the core. In Fig. 2.5c it is clear that if the faces and the core are to remain in contact, the faces are being called upon to bend to an infinite curvature at the centre of the beam. This is not possible; instead the faces bend locally some distance either side of the centre-line of the beam as in Fig. 2.5d, smoothing out the sharp discontinuity in the shear deflection curve. In doing so, the faces reduce the shear deflection at the expense of introducing additional bending moments and shear forces into the faces.

In many practical sandwiches, especially those with thin faces, the effect is small. When the faces are thicker (as with asbestos cement or plywood for example) and the core is weak the effect may be significant. It may also be necessary to take the effect into account when performing tests on sandwich beams to determine the shear modulus of the core.

To discuss the nature of this interaction between the bending stiffness of the faces and the shear stiffness of the core, consider first a sandwich with a core which is rigid in shear ($G = \infty$) and with a load q_1 per unit length. A deflection w_1 occurs in accordance with ordinary bending theory. This deflection is associated with a bending moment M_1 and a shear force Q_1, the latter being given by:

$$-Q_1 = Dw_1''' = E(I-I_f)w_1''' + EI_f w_1'''. \qquad (2.18)$$

Primes denote differentiation with respect to x. E is the modulus of elasticity of the faces, I is the second moment of area of the faces about the centroid of the sandwich and I_f is the sum of the second moments of area of the faces about their own centroids. The contribution of the core to D is neglected.

$$I = \frac{bt^3}{6} + \frac{btd^2}{2}, \qquad (2.19a)$$

$$I_f = \frac{bt^3}{6}. \qquad (2.19b)$$

The first term on the right-hand side of equation (2.18) represents the shear force carried by the beam as a whole, supposing the faces to undergo only uniform extensions or contractions without bending locally. In this state the shear stress τ is uniform across the thickness of the core (c) and diminishes linearly to zero across the thickness of each face (Fig. 2.4c). The first term may therefore be replaced by $-bd\tau$, where τ is the shear stress in the core:

$$-Q_1 = -bd\tau + EI_f w_1'''. \qquad (2.20)$$

Also,
$$q_1 = -Q_1', \quad Q_1 = M_1', \quad M_1 = -Dw_1''. \qquad (2.21)$$

As a result of the shear stress τ, the core undergoes a shear strain $\gamma = \tau/G$, which corresponds to an additional transverse beam deflection w_2. The faces must share this extra deflection and, in order to do so, they must be subjected to additional distributed loads q_2, shear forces Q_2 and bending moments M_2 such that

$$q_2 = -Q_2'; \quad Q_2 = M_2'; \quad M_2 = -EI_f w_2''. \quad (2.22)$$

The total loads, shear forces, bending moments and deflections are now:

$$q = q_1 + q_2, \quad (2.23a)$$
$$Q = Q_1 + Q_2, \quad (2.23b)$$
$$M = M_1 + M_2, \quad (2.23c)$$
$$w = w_1 + w_2. \quad (2.23d)$$

In different words it may be said that a sandwich beam with a total load q undergoes two distinct sets of displacements, w_1 and w_2. The first represents the ordinary bending deflection associated with a shear force Q_1 which is shared between the faces and the core (equation (2.20)). The second represents the shear deflection of the core due to Q_1. The faces participate in this extra deflection by bending about their own axes. In doing so, they support an extra shear force Q_2. The sum of Q_1 and Q_2 is the shear force applied to the beam.

The relationship between the core shear strain γ and the additional deflection may be obtained from Fig. 2.6, as in Section 2.3. From equation (2.15a),

$$\gamma = \frac{d}{c} \cdot w_2'. \quad (2.24a)$$

Furthermore, the core shear stress is

$$\tau = \frac{d}{c} \cdot G w_2'. \quad (2.24b)$$

Substitution for τ in equation (2.20) yields:

$$-Q_1 = -AGw_2' + EI_f w_1''', \quad (2.25)$$

where $A = bd^2/c$.

Rearrangement of equation (2.25) and the substitution $Q_1 = -Dw_1'''$ gives

$$w_2' = -\frac{D}{AG}\left(1-\frac{I_f}{I}\right)w_1''' = +\frac{Q_1}{AG}\left(1-\frac{I_f}{I}\right). \qquad (2.26)$$

The total shear force is:

$$Q = Q_1 + Q_2 = Q_1 - EI_f w_2'''.$$

Substitution for w_2''' from equation (2.26) provides a differential equation† for Q_1:

$$Q_1'' - a^2 Q_1 = -a^2 Q, \qquad (2.27a)$$

where

$$a^2 = \frac{AG}{EI_f(1-I_f/I)}. \qquad (2.27b)$$

In any particular problem in which Q is a given function of x, this equation can be solved for Q_1. The quantities M_1, w_1, q_1 may be obtained by integration and differentiation. The slope w_2' may be found from equation (2.26) and the related quantities M_2, w_2, q_2 again obtained by integration and differentiation.

The quantity a^2 represents essentially the ratio of the core shear stiffness to the local bending stiffness of the faces. The extent to which the faces modify the shear deformation of the core depends partly on a^2 and partly on the span, the effect being smaller at large values of a^2 and at large spans.

Equation 2.26 may be rewritten, more compactly,

$$EI_f w_2' = \frac{Q_1}{a^2}. \qquad (2.28)$$

In the foregoing analysis it was tacitly assumed that during the secondary deformation w_2, the points d and a (Fig. 2.6) remained in the same vertical plane. This is true provided that either the beam is loaded and supported in a symmetrical manner, or some suitable physical restraint is applied to the beam (for example,

† A similar equation was obtained by Norris[16.2] by different reasoning.

when at least one end of the beam is fixed, or when the ends of the faces are pinned to a rigid abutment.)

If these conditions are not fulfilled, equations (2.24) must be written in a more general form:

$$\gamma = \frac{d}{c} w'_2 + \gamma_0, \tag{2.24c}$$

$$\tau = \frac{d}{c} G w'_2 + G \gamma_0. \tag{2.24d}$$

Here γ_0 is a constant, which is equal to the shear strain when $w'_2 = 0$. With the introduction of γ_0, equations (2.25), (2.26) and (2.28) are modified as shown below. Equations (2.27) are unaltered.

$$-Q_1 = -AGw'_2 + EI_f w'''_1 - bdG\gamma_0, \tag{2.25a}$$

$$w'_2 = -\frac{D}{AG}\left(1 - \frac{I_f}{I}\right) w'''_1 - \frac{c}{d}\gamma_0 = \frac{Q_1}{AG}\left(1 - \frac{I_f}{I}\right) - \frac{c}{d}\gamma_0, \tag{2.26a}$$

$$EI_f w'_2 = \frac{Q_1}{a^2} - \frac{c}{d} \gamma_0 EI_f. \tag{2.28a}$$

Very few practical problems involve non-zero values of γ_0.

2.6. Simply-supported Beam with Central Point Load W (Antiplane Core and Thick Faces)

By symmetry it is necessary to consider only the right-hand half of the beam, which is in effect a cantilever of length $L/2$, with an overhang L_1 (Fig. 2.9). Furthermore, γ_0 is zero.

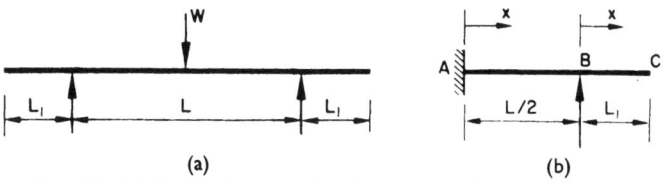

FIG. 2.9. (a) Beam with central point load. (b) Right-hand half of the beam.

In the part AB, with x measured from A, the shear force is $-W/2$; the solution of equation (2.27) with this value allocated to Q is

$$-Q_1 = C_1 \cosh ax + C_2 \sinh ax + \frac{W}{2}. \tag{2.29}$$

By successive integration,

$$EIw_1 = \frac{C_1}{a^3} \sinh ax + \frac{C_2}{a^3} \cosh ax + \frac{Wx^3}{12} + C_3 x^2 + C_4 x + C_5. \tag{2.30}$$

Equations (2.28) and (2.29) together provide an expression for w_2' which may be integrated once:

$$-EI_f w_2 = \frac{C_1}{a^3} \sinh ax + \frac{C_2}{a^3} \cosh ax + \frac{W}{2a^2} x + C_6. \tag{2.31}$$

Five simple boundary conditions are available in AB to provide relationships between the six constants of integration, $C_1 - C_6$.

(i) $\underline{x = 0, \quad w_1 = 0}$ (Arbitrary)

$$C_5 + \frac{C_2}{a^3} = 0.$$

(ii) $\underline{x = 0, \quad w_1' = 0}$ (Symmetry)

$$\frac{C_1}{a^2} + C_4 = 0.$$

(iii) $\underline{x = 0, \quad w_1''' = 0}$ (Symmetry)

$$C_1 + \frac{W}{2} = 0.$$

(iv) $x = 0, \quad M = \dfrac{WL}{4}$

By definition, $-M = EIw_1'' + EI_f w_2''$.

Hence $\quad -\dfrac{WL}{4} = 2C_3.$

(v) $\underline{x = 0, \quad w_2 = 0}$ (Arbitrary)

$$\frac{C_2}{a^3} + C_6 = 0.$$

As a result, the constants may be expressed as follows, C_2 being still unknown:

$$C_1 = -\frac{W}{2}; \quad C_3 = -\frac{WL}{8}; \quad C_4 = +\frac{W}{2a^2}; \quad C_5 = C_6 = -\frac{C_2}{a^3}. \tag{2.32}$$

In the part BC, with x measured from B, the total shear force is zero. Equations (2.29) and (2.30) are therefore valid provided the terms containing W are suppressed and new constants $B_1 - B_6$ are introduced to replace $C_1 - C_6$.

There are four simple boundary conditions, as follows:

(vi) $\underline{x = 0, \quad w_1 = 0}$ (Arbitrary)
$$B_5 + \frac{B_2}{a^3} = 0.$$

(vii) $\underline{x = 0, \quad w_2 = 0}$ (Arbitrary)
$$B_6 + \frac{B_2}{a^3} = 0.$$

(viii) $\underline{x = L_1, \quad w_1'' = 0}$
$$\frac{B_1}{a}\sinh aL_1 + \frac{B_2}{a}\cosh aL_1 + 2B_3 = 0.$$

(ix) $\underline{x = L_1, \quad w_2'' = 0}$
$$\frac{B_1}{a}\sinh aL_1 + \frac{B_2}{a}\cosh aL_1 = 0.$$

The last two conditions arise because M_1 and M_2 are assumed to vanish separately at the free end. This is true provided the ends of the faces are free to rotate, and are not attached to a rigid end diaphragm. The following results are obtained from these boundary conditions:

$$B_2 = -B_1 \tanh aL_1; \quad B_3 = 0; \quad B_5 = B_6 = \frac{B_1}{a^2}\tanh aL_1. \tag{2.33}$$

It remains to establish continuity at B. It is evident that w_1', w_2', w_1'' and w_2'' must be continuous; also, from equation (2.26) w_1''' and w_1^{iv} must be continuous in consequence. However, the only three conditions which provide independent equations are the continuity of w_1', w_2' and w_1''.

(x) w_1' continuous at B

$$\frac{C_1}{a^2}\cosh\frac{aL}{2}+\frac{C_2}{a^2}\sinh\frac{aL}{2}+\frac{WL^2}{16}+C_3L+C_4 = \frac{B_1}{a^2}+B_4.$$

(xi) w_2' continuous at B

$$C_1\cosh\frac{aL}{2}+C_2\sinh\frac{aL}{2}+\frac{W}{2} = B_1.$$

(xii) w_1'' continuous at B

$$C_1\sinh\frac{aL}{2}+C_2\cosh\frac{aL}{2}+\left(\frac{WL}{4}+2C_3\right)a = B_2+2B_3a.$$

B_2, B_3, C_1, C_3 may be eliminated by the use of expressions (2.32) and (2.33); conditions (xi) and (xii) provide two equations to be solved for C_2 and B_1. Only C_2 is of immediate interest:

$$C_2 = \beta_1\frac{W}{2}, \tag{2.34}$$

where

$$\beta_1 = \frac{\sinh\theta-(1-\cosh\theta)\tanh\phi}{\sinh\theta\tanh\phi+\cosh\theta}; \quad \theta = \frac{aL}{2}; \quad \phi = aL_1.$$

(Condition (x) may be used to find B_4 if necessary.)

The values of C_1-C_6, now known, may be written into equations (2.30) and (2.31). The total deflection, w, may therefore be expressed as a function of x in the region AB. Equation (2.24b) may be used to express the core shear stress τ as a function of x in the same region. Furthermore, the maximum direct stress in the faces at any section is:

$$\sigma = M_1\frac{c+2t}{2I}+M_2\frac{t}{2I_f}. \tag{2.35}$$

Double differentiation of equations (2.30) and (2.31) yields M_1 and M_2, and hence σ, as a function of x in AB. The results of these operations are summarized below.

$$w = -\frac{Wx^2L}{24EI}\left(3-\frac{2x}{L}\right) - \frac{WL}{4AG}\left(1-\frac{I_f}{I}\right)^2$$
$$\times \left\{\frac{2x}{L} - \frac{2}{aL}[\sinh ax + \beta_1(1-\cosh ax)]\right\}, \quad (2.36a)$$

$$\tau = -\frac{W}{2bd}\left(1-\frac{I_f}{I}\right)\{1-\cosh ax + \beta_1 \sinh ax\}, \quad (2.36b)$$

$$\sigma = \frac{WL}{4}\left\{\left[\left(1-\frac{2x}{L}\right) - \frac{2}{aL}(\beta_1 \cosh ax - \sinh ax)\right]\frac{c+2t}{2I}\right.$$
$$\left. + \frac{2}{aL}(\beta_1 \cosh ax - \sinh ax)\frac{t}{2I_f}\right\}. \quad (2.36c)$$

The peak values of w, τ and σ occur at $x=L/2$, $x=(1/a)\tanh^{-1}\beta_1$ and $x=0$, respectively:

$$w_{\max} = -\frac{WL^3}{48EI} - \frac{WL}{4AG}\left(1-\frac{I_f}{I}\right)^2 \psi_1, \quad (2.37a)$$

$$\tau_{\max} = -\frac{W}{2bd}\left(1-\frac{I_f}{I}\right)\psi_2, \quad (2.37b)$$

$$\sigma_{\max} = \frac{WL}{4}\left\{\frac{c+2t}{2I}\psi_3 + \frac{t}{2I_f}(1-\psi_3)\right\}, \quad (2.37c)$$

where $\psi_1 = 1 - \dfrac{\sinh\theta + \beta_1(1-\cosh\theta)}{\theta}$, \quad (2.38a)

$$\psi_2 = 1 - \sqrt{(1-\beta_1^2)}, \quad (2.38b)$$

$$\psi_3 = 1 - \frac{\beta_1}{\theta}. \quad (2.38c)$$

To facilitate the use of equations (2.37) in particular problems, Fig. 2.10 provides curves of ψ_1, ψ_2, ψ_3 plotted against θ for the two extreme cases, $\phi = 0$ (no overhang) and $\phi = \infty$ (infinite overhang). The value of θ itself may be determined quickly from Fig. 2.11, which has been plotted with the aid of equation (2.27b):

$$\theta = \frac{aL}{2} = \frac{L}{2}\left[\frac{AG}{EI_f\left(1-\dfrac{I_f}{I}\right)}\right]^{\frac{1}{2}} = \frac{L}{c}\left[\frac{G}{2E}\frac{c}{t}\left(1+\frac{3d^2}{t^2}\right)\right]^{\frac{1}{2}}.$$

(2.39)

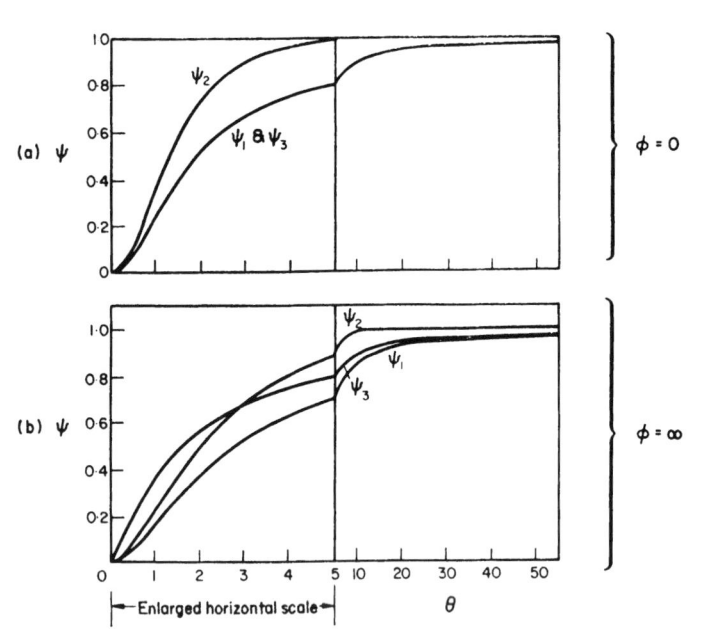

Fig. 2.10. Values of ψ_1, ψ_2, ψ_3 in equation (2.38).

Equations (2.37) represent the behaviour of a sandwich whose properties lie somewhere between those extreme cases indicated in Table 2.1.

In case (i) the core vanishes and the faces act as two independent beams. Ordinary bending theory confirms the expressions for maximum deflection and direct stress.

In case (ii) the core is rigid in shear and the sandwich acts as a single composite beam. Again, ordinary bending theory confirms the results shown.

In case (iii) the faces are thin membranes. The maximum deflection agrees with the value of Δ in Section 2.3 (for $c = d$) and

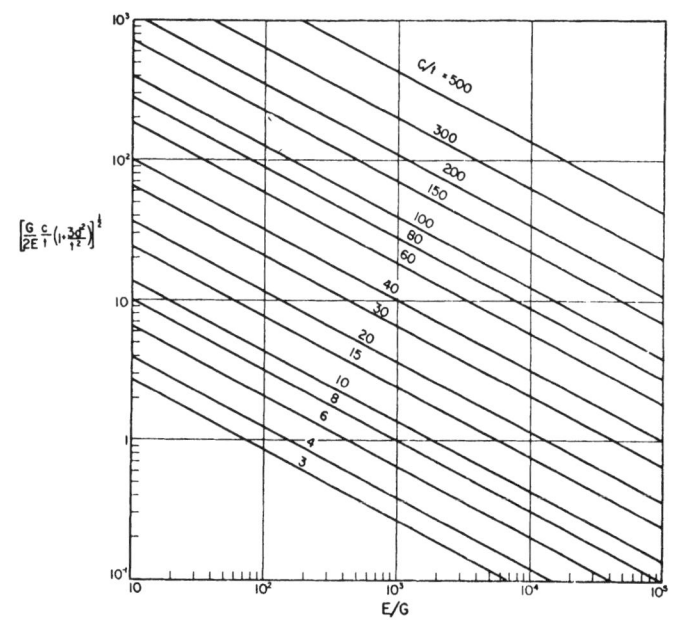

FIG. 2.11. Graph for the determination of θ (equation (2.39)):
$$\theta = \frac{L}{c}\left[\frac{G}{2E}\frac{c}{t}\left(1+\frac{3d^2}{t^2}\right)\right]^{\frac{1}{2}}.$$

the maximum shear stress agrees with equation (2.13b) (for $Q = -W/2$). In the expression for the maximum direct stress the first term represents the membrane stress in the faces, as predicted by ordinary bending theory. The second term describes the local bending stress, which becomes infinite as t tends to zero; it should

TABLE 2.1. EXTREME CASES FOR A SIMPLY-SUPPORTED BEAM
WITH A CENTRAL POINT LOAD

	Limiting values		
	(i) $G \to 0$	(ii) $G \to \infty$	(iii) $t \to 0$
a	0	∞	∞
θ	0	∞	∞
β_1	0	1	1
ψ_1	$\dfrac{\theta^3}{3}$	1	1
ψ_2	0	1	1
ψ_3	0	1	$1 - \dfrac{1}{\theta}$
w_{max}	$-\dfrac{WL^3}{48EI_f}$	$-\dfrac{WL^3}{48EI}$	$-\dfrac{WL^3}{48EI} - \dfrac{WL}{4bdG}$
τ_{max}	0	$-\dfrac{W}{2bd}\left(1 - \dfrac{I_f}{I}\right)$	$-\dfrac{W}{2bd}$
σ_{max}	$\dfrac{WL}{4} \cdot \dfrac{t}{2I_f}$	$\dfrac{WL}{4} \cdot \dfrac{c+2t}{2I}$	$\dfrac{WL}{4} \cdot \dfrac{c+2t}{2I}$ $+ \dfrac{WL}{4} \cdot \dfrac{t}{2I_f} \cdot \dfrac{1}{\theta}$

be noted that this expression represents conditions at the point of infinite curvature, c, in Fig. 2.5c.

The functions ψ diverge significantly from unity only when θ is small (less than 20, say). This happens when the span to depth ratio (L/c) is small and when the shear stiffness of the core is low in relation to the flexural rigidity of the faces (see equation 2.39). In practice this means a weak core (such as expanded polystyrene) with thick low-strength faces (such as asbestos cement). It is worth noting, however, that the further the functions ψ depart from unity, the more the sandwich behaves as two separate beams (the faces) unconnected by the core. Since the whole purpose of sandwich construction is to minimize this effect by having a core rigid enough to make one face work principally in compression, the other in tension, it is clear that values of θ much smaller than 20 signify an inefficient structure.

2.7. Simply-supported Beam with Uniformly Distributed Load (Antiplane Core and Thick Faces)

The right-hand half of the beam is shown in Fig. 2.12 as a cantilever AB of length $L/2$, with an unloaded overhang BC of length L_1. By symmetry, γ_0 is zero.

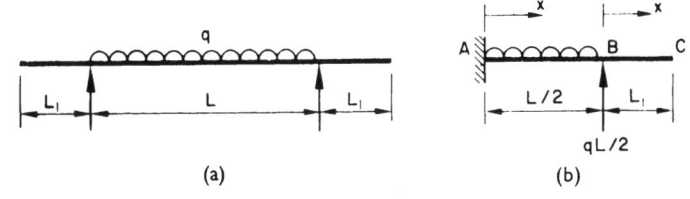

FIG. 2.12. (a) Beam with uniformly distributed load. (b) Right-hand half of the beam.

In the part AB, with x measured from A, the shear force is $-qx$. The solution of equation (2.27) with this value allocated to Q is

$$-Q_1 = C_1 \cosh ax + C_2 \sinh ax + qx. \qquad (2.40)$$

By successive integration,

$$EIw_1 = \frac{C_1}{a^3} \sinh ax + \frac{C_2}{a^3} \cosh ax + \frac{qx^4}{24} + C_3 x^2 + C_4 x + C_5. \qquad (2.41)$$

Equations (2.28) and (2.40) together provide an expression for w_2' which may be integrated once:

$$-EI_f w_2 = \frac{C_1}{a^3} \sinh ax + \frac{C_2}{a^3} \cosh ax + \frac{qx^2}{2a^2} + C_6. \qquad (2.42)$$

Equations (2.41) and (2.42) are valid for BC, with x measured from B provided the terms containing q are suppressed and the constants C_1-C_6 are replaced by new constants B_1-B_6.

The boundary conditions and the requirements for continuity at B are the same as those marked (i)—(xii) in the previous section, except that in (iv) $WL/4$ is replaced by $qL^2/8$. The process of evaluating the constants is also similar to that adopted in the previous section, and the results are listed below:

$$C_1 = 0; \quad C_3 = -\frac{qL^2}{16} + \frac{q}{2a^2}; \quad C_4 = 0; \quad C_5 = C_6 = -\frac{C_2}{a^3},$$
(2.43a)

$$B_1 = C_2 \sinh \frac{aL}{2} + \frac{qL}{2}; \quad B_2 = -B_1 \tanh aL_1; \quad B_3 = 0;$$

$$B_4 = -\frac{qL^3}{24}, \tag{2.43b}$$

$$C_2 = -\beta_2 \frac{qL}{2}, \tag{2.43c}$$

where

$$\beta_2 = \frac{1/\theta + \tanh \phi}{\cosh \theta + \sinh \theta \tanh \phi}; \quad \theta = \frac{aL}{2}; \quad \phi = aL_1.$$

The formulae for the deflection, core shear stress and direct stress in the faces for any point in AB are analogous with equations (2.36):

$$w = -\frac{qx^2L^2}{48EI}\left(3 - \frac{2x^2}{L^2}\right) - \frac{q}{AG}\left(1 - \frac{I_f}{I}\right)^2 \left\{\frac{x^2}{2} + \frac{\beta_2 L^2}{4\theta}(1-\cosh ax)\right\},$$
(2.44a)

$$\tau = -\frac{qL}{2bd}\left(1 - \frac{I_f}{I}\right)\left\{\frac{2x}{L} - \beta_2 \sinh ax\right\}, \tag{2.44b}$$

$$\sigma = \frac{qL^2}{8}\left\{\left[\left(1 - \frac{4x^2}{L^2}\right) - \frac{2}{\theta^2}(1-\beta_2\theta \cosh ax)\right]\frac{c+2t}{2I}\right.$$
$$\left. + \frac{2}{\theta^2}(1 - \beta_2\theta \cosh ax)\frac{t}{2I_f}\right\}. \tag{2.44c}$$

The maximum values of w, τ and σ occur, respectively, at

$$x = \frac{L}{2}, \quad x = \frac{1}{a}\cosh^{-1}\left(\frac{1}{\theta\beta_2}\right), \quad x = 0.$$

$$w_{max} = -\frac{5qL^4}{384EI} - \frac{qL^2}{8AG}\left(1 - \frac{I_f}{I}\right)^2 \psi_4, \tag{2.45a}$$

$$\tau_{max} = -\frac{qL}{2bd}\left(1 - \frac{I_f}{I}\right)\psi_5, \tag{2.45b}$$

$$\sigma_{max} = \frac{qL^2}{8}\left\{\frac{c+2t}{2I}\psi_6 + \frac{t}{2I_f}(1-\psi_6)\right\}, \tag{2.45c}$$

where $\psi_4 = 1 + \frac{2\beta_2}{\theta}(1-\cosh\theta)$, (2.46a)

$$\psi_5 = \frac{2x_1}{L} - \beta_2 \sinh ax_1, \quad \text{where} \quad \cosh ax_1 = \frac{1}{\beta_2\theta}, \tag{2.46b}$$

$$\psi_6 = 1 - \frac{2}{\theta^2}(1-\beta_2\theta). \tag{2.46c}$$

The equations (2.45) reduce to simple forms in the three extreme cases listed in Table 2.2. The functions ψ_4, ψ_5, ψ_6 are plotted in

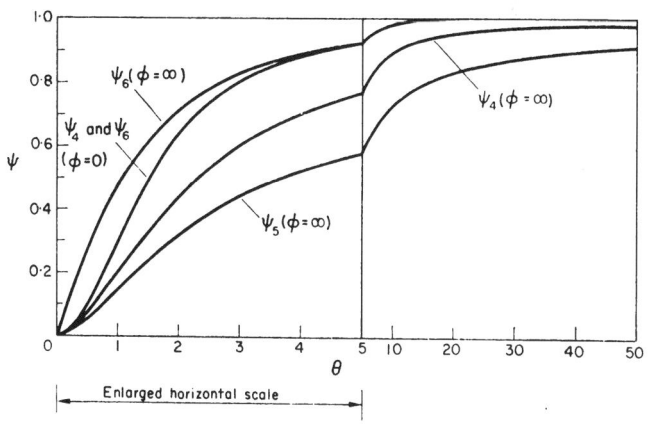

Fig. 2.13. Values of ψ_4, ψ_5, ψ_6 in equation (2.46). (When $\phi = 0$, ψ_5 is equal to ψ_3 in Fig. 2.10a.)

TABLE 2.2. EXTREME CASES FOR A SIMPLY-SUPPORTED BEAM WITH UNIFORMLY-DISTRIBUTED LOAD

	Limiting values		
	(i) $G \to 0$	(ii) $G \to \infty$	(iii) $t \to 0$
a	0	∞	∞
θ	0	∞	∞
β_2	$\dfrac{1}{\theta}\left(1-\dfrac{\theta^2}{2}\right) \to \infty$	$\dfrac{2 \tanh \phi}{e^\theta(1+\tanh \phi)} \to 0$	$\dfrac{2 \tanh \phi}{e^\theta(1+\tanh \phi)} \to 0$
ψ_4	$\dfrac{5}{12}\theta^2$	1	1
ψ_5	0	1	1
ψ_6	0	1	1
w_{max}	$-\dfrac{5}{384}\dfrac{qL^4}{EI_f}$	$-\dfrac{5}{384}\dfrac{qL^4}{EI}$	$-\dfrac{5}{384}\dfrac{qL^4}{EI}-\dfrac{qL^2}{8bdG}$
τ_{max}	0	$-\dfrac{qL}{2bd}\left(1-\dfrac{I_f}{I}\right)$	$-\dfrac{qL}{2bd}$
σ_{max}	$\dfrac{qL^2}{8}\dfrac{t}{2I_f}$	$\dfrac{qL^2}{8}\dfrac{c+2t}{2I}$	—

Fig. 2.13 in terms of θ for the two extreme cases $L_1 = 0$ and $L_1 = \infty$. As before, θ can be found from Fig. 2.11.

2.8. Beam with Four-point Loading (Antiplane Core and Thick Faces)

An ordinary homogeneous beam with four-point loading is illustrated in Fig. 2.14a. The central region is subjected to a constant bending moment $-WL_b$ and it therefore bends into a circular arc of curvature $+WL_b/EI$. This fact is often used to determine the flexural rigidity EI of a simple beam, usually by measuring the deflections of three different points in the central region.

The method is also used to determine the flexural rigidity of sandwich beams, on the assumption that because there is no shear force in the central region, there can be no shear deflection there and the deformation is one of pure bending. This is true for sandwich beams with thin faces and stiff cores. When the faces are thick and the core is weak, however, this simple view is no longer valid and the central region no longer bends in the arc of a circle.

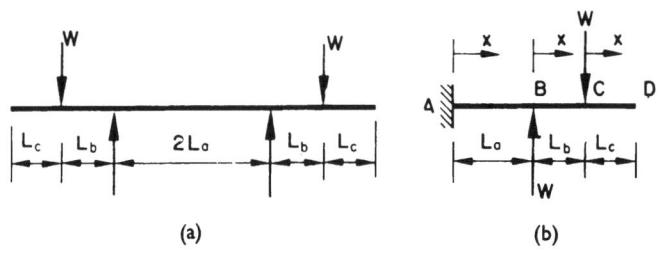

Fig. 2.14. (a) Beam with four-point loading. (b) Right-hand half of the beam.

The effect may be analysed by the method of Section 2.6, applied to the right-hand half of the beam (Fig. 2.14b). By symmetry, γ_0 is zero.

The shear force Q is equal to $+W$ in BC and zero in AB and BC. The solution of equation (2.27) for BC is

$$-Q_1 = B_1 \cosh ax + B_2 \sinh ax + W. \qquad (2.47)$$

By successive integration the primary bending deformation is given by

$$EIw_1 = \frac{B_1}{a^3} \sinh ax + \frac{B_2}{a^3} \cosh ax + B_3 x^2 + B_4 x + B_5 - \frac{Wx^3}{6}. \qquad (2.48)$$

Equations (2.28) and (2.47) together provide an expression for w_2' which may be integrated at once:

$$-EI_f w_2 = \frac{B_1}{a^3} \sinh ax + \frac{B_2}{a^3} \cosh ax + B_6 - \frac{Wx}{a^2}. \qquad (2.49)$$

Similar equations are applicable to regions AB and CD except that the constants $B_1 - B_6$ are replaced by other constants $A_1 - A_6$ and $C_1 - C_6$, respectively, and the terms containing W vanish.

Boundary conditions are as follows:

AB $x = 0$, $w_1 = w_1' = w_1''' = w_2 = 0$, $M = -W \cdot L_b$.

Hence, $A_1 = A_4 = 0$, $A_3 = W \cdot L_b/2$, $A_5 = A_6 = -A_2/a^3$.

BC $x = 0$, $w_1 = w_2 = 0$, $M = -W \cdot L_b$.

Hence $B_3 = W \cdot L_b/2$, $B_5 = B_6 = -B_2/a^3$.

CD $x = 0$, $w_1 = w_2 = 0$; $x = L_c$, $w_1'' = w_2'' = 0$.

Hence $C_2 = -C_1 \tanh aL_c$, $C_3 = 0$, $C_5 = C_6 = -C_2/a^3$.

At B and C

w_1', w_2', w_1'' are continuous.

In all there are eighteen equations and eighteen constants to be determined. It is not difficult to show that:

$$A_2 = \beta_3 W, \tag{2.50}$$

where

$$\beta_3 = \frac{\sinh aL_c - \sinh a(L_b + L_c)}{\cosh a(L_a + L_b + L_c)}.$$

The total deflection at any point x in AB may be found from the equations analogous with (2.48) and (2.49), noting that A_1 and A_4 are zero, A_3 is $W \cdot L_b/2$, and A_2 is given by equation (2.50). After some manipulation the following expression results:

$$w = \frac{W \cdot L_b}{2} \frac{x^2}{EI} + \frac{\beta_3 W}{a^3 EI_f} (1 - \cosh ax) \left(1 - \frac{I_f}{I}\right). \tag{2.51}$$

It is now convenient to review the analysis which is usually applied to a beam under four-point loading in order to determine the flexural rigidity EI. Figure 2.15 shows the central portion of the beam, EAF, with E and F equidistant from A. Under the influence of a constant bending moment, M, this part bends

SANDWICH BEAMS

into an arc of a circle of centre O and curvature $1/R = M/EI$. Deflections are measured at E, A and F so that the displacement, Δ, of A with respect of EF can be found. From the right-angled triangle OBF,

$$AF^2 \doteq BF^2 = R^2 - (R-\Delta)^2.$$

Because Δ is small in comparison with R, this can be simplified to $AF^2 = 2R\Delta$. Hence:

$$\frac{1}{EI} = -\frac{1}{MR} = -\frac{2\Delta}{M \cdot AF^2}. \tag{2.52}$$

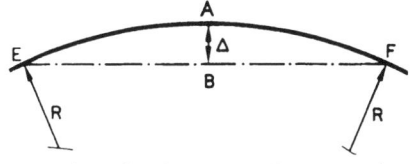

FIG. 2.15. Beam bent in the shape of an arc of a circle.

If this same procedure is applied to the region AB of the sandwich beam of Fig. 2.14, with $AF = x$, $M = -W \cdot L_b$ and $\Delta = w - (w)_{x=0} = w$, then equation (2.52) gives not $1/EI$ but $1/(EI)'$, the reciprocal of the *apparent* (and erroneous) flexural rigidity:

$$\frac{1}{(EI)'} = \frac{2w}{W \cdot L_b x^2}, \tag{2.53}$$

where w is given by equation (2.51).

The fractional error in $1/EI$ is therefore

$$\frac{1/(EI)' - 1/EI}{1/EI} = \frac{EI}{(EI)'} - 1 = \frac{2\beta_3}{L_b a^3 x^2} \frac{I}{I_f} (1 - \cosh ax) \left(1 - \frac{I_f}{I}\right)$$

$$= \frac{2\beta_3}{aL_b} \left(\frac{1 - \cosh ax}{a^2 x^2}\right) \left(\frac{I}{I_f} - 1\right). \tag{2.54}$$

This equation may be used to estimate the error which is likely to occur in the measurement of $1/EI$ in any particular experiment.

If the error is small, the fractional errors in $1/EI$ and EI are similar but not identical.

There are too many variables for this result to be expressed conveniently in graphical form, but some general conclusions are possible.

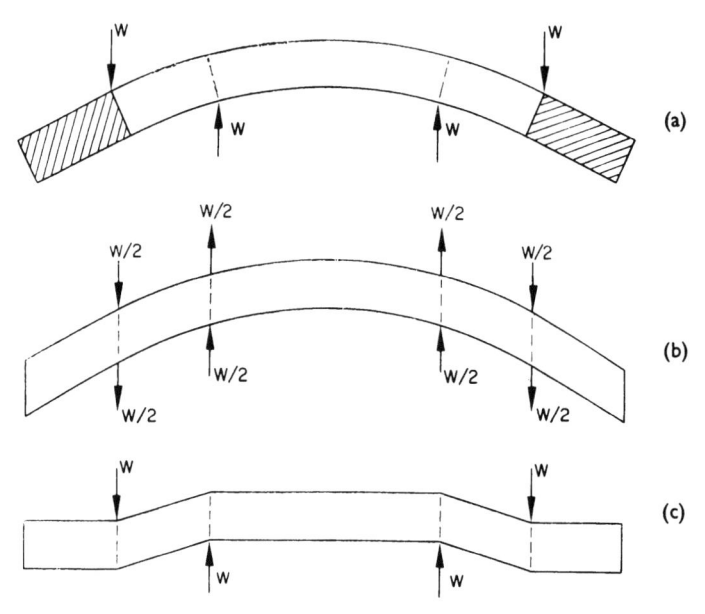

FIG. 2.16. Deformations of a sandwich beam with four-point loading. (a) With rigid inserts (shaded). (b) With weak core. (c) In pure shear.

For example, if $L_c \to \infty$, then $\beta_3 \to 0$, the fractional error (equation (2.54)) vanishes and the experiment gives accurate values of EI in all circumstances. This is because the large amount of core material in the infinite overhang acts like a rigid insert, preventing the faces from sliding over each other (Fig. 2.16a). This suggests that one way of eliminating errors in tests of this kind would be to glue rigid inserts into the overhanging parts, in place of the core. Some of the simplicity of the method is then lost, however, because tests can no longer be performed on unmodified sample beams.

If $G \to \infty$, then $a \to \infty$, $\beta_3 \to 0$ and the fractional error vanishes. In this case the sandwich acts as a single composite beam without shear deformation.

If $G \to 0$, then $a \to 0$, $\beta_3 \to -aL_b$ and the fractional error becomes $(I/I_f - 1)$. This is consistent with the faces acting as two separate beams with a total flexural rigidity of EI_f. The deformation is illustrated in Fig. 2.16b, which clearly shows how the faces slide over each other, even in the central portion of the beam.

If $t \to 0$, then $I_f \to 0$, $a \to \infty$, $\beta_3 \to 0$ and the fractional error in $1/EI$ again vanishes. This is because the shear deformations in this case have no effect at all in the central region of the beam. The shear deformations shown in Fig. 2.16c illustrate this point.

2.9. Sandwich Beams with Faces of Unequal Thickness (Antiplane Core)

Few alterations are necessary to the results of the previous sections when the faces are of unequal thickness. Consider the cross-section of a sandwich beam in which the thicknesses of the upper and lower faces are t_1, t_2, respectively (Fig. 2.17). As before, d represents the distance between the centroids of the upper and lower faces. The second moment of area of the sandwich as a

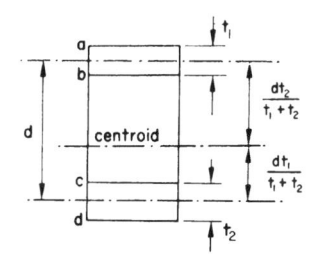

FIG. 2.17. Dimensions of the cross-section of a sandwich with faces of unequal thickness.

whole (I) and the sum of the second moments of area of the faces bending separately about their own centroidal axes (I_f) are:

$$I = \frac{bd^2 t_1 t_2}{t_1+t_2} + \frac{b}{12}(t_1^3+t_2^3), \tag{2.55a}$$

$$I_f = \frac{b}{12}(t_1^3+t_2^3). \tag{2.55b}$$

These replace equations (2.19).

The results of Section 2.1 are modified in detail when the faces are unequal, but it is useful to note that equation (2.13b), for the core shear stress when the faces are thin, is unaltered.

Apart from the two major changes discussed below, Sections 2.2–2.8 are unchanged. In particular, equations (2.18) and (2.24a) remain the same.

The expression on the extreme right-hand side of equation (2.39) is, of course, no longer valid. Instead, equations (2.55) must be used for I and I_f, and AG is again represented by bd^2G/c.

The second change concerns equation (2.35) for the direct stress σ in the faces, which must now be written:

$$\sigma = M_1\alpha_1 + M_2\alpha_2, \tag{2.56}$$

TABLE 2.3.

Level	α_1	α_2
a	$-\dfrac{1}{I}\left(\dfrac{dt_2}{t_1+t_2}+\dfrac{t_1}{2}\right)$	$-\dfrac{1}{I_f}\dfrac{t_1}{2}$
b	$-\dfrac{1}{I}\left(\dfrac{dt_2}{t_1+t_2}-\dfrac{t_1}{2}\right)$	$+\dfrac{1}{I_f}\dfrac{t_1}{2}$
c	$+\dfrac{1}{I}\left(\dfrac{dt_1}{t_1+t_2}-\dfrac{t_2}{2}\right)$	$-\dfrac{1}{I_f}\dfrac{t_2}{2}$
d	$+\dfrac{1}{I}\left(\dfrac{dt_1}{t_1+t_2}+\dfrac{t_2}{2}\right)$	$+\dfrac{1}{I_f}\dfrac{t_2}{2}$

where α_1, α_2 take different values according to the level at which the stress is measured. For example, the four critical levels are marked a, b, c, d in Fig. 2.17, for which the values of α are given in Table 2.3. Tensile stresses and sagging bending moments are positive.

The quantities α_1 and α_2 replace the quantities $(c+2t)/2I$ and $t/2I_f$ respectively in equations (2.36c), (2.37c), (2.44c) and (2.45c).

2.10. Sandwich Beams in which the Modulus of Elasticity of the Core Parallel with the Axis is not Small (Faces of Equal Thickness)

When the modulus of elasticity, E_c, of the core is not small, it is necessary to use the full expression for the flexural rigidity, D, of the sandwich (equation (2.2)). Furthermore, because condition (2.11) is not satisfied, the shear stress τ and the shear strain γ are no longer constant over the depth of the core. Equation (2.10) is valid, but not the simple form (2.12); the core is no longer antiplane.

Figure 2.18 shows an elevation of a short length of sandwich beam, undergoing shear deformation of the core. The plane section ACB has distorted into the curve $A'CB'$, the typical point P having moved a distance u to the right. The shear strain γ at

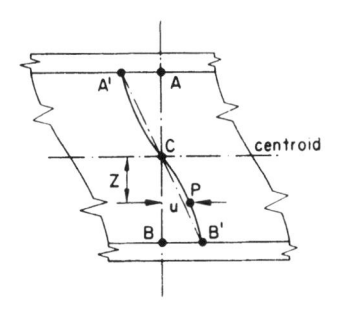

FIG. 2.18. Shear deformation of sandwich beam with stiff core.

P is therefore du/dz and the shear stress is

$$\tau = G\frac{du}{dz}. \tag{2.57}$$

Equations (2.10) and (2.57) may be combined and integrated to yield an expression for u:

$$u = \frac{Q}{GD}\left\{\frac{E_f t dz}{2} + \frac{E_c}{2}\left(\frac{c^2 z}{4} - \frac{z^3}{3}\right)\right\}. \tag{2.58}$$

There is no constant of integration because u and z vanish simultaneously. The displacements AA' and BB' are obtained by writing $z = \pm c/2$.

$$BB' = \frac{Q}{GD}\left(\frac{E_f}{4} tdc + \frac{E_c c^3}{24}\right). \tag{2.59}$$

The maximum shear stress is obtained by writing $z = 0$ in equation (2.10).

$$\tau_{\max} = \frac{Q}{D}\left(\frac{E_f t d}{2} + \frac{E_c c^2}{8}\right). \tag{2.60}$$

Suppose now that the real core is replaced by a true antiplane core with a shear modulus G', different from G, but that the correct value of D is retained. The foregoing analysis may be repeated except that G is replaced by G' and E_c vanishes. The value of G' may be chosen so that the section ACB now deforms into the *straight line* $A'CB'$, and the horizontal displacement of the lower edge of the core is

$$BB' = \frac{Q}{G'D}\left(\frac{E_f t d c}{4}\right). \tag{2.61}$$

Because G' has been chosen so that equations (2.59) and (2.60) give the same result for BB', then the antiplane core (G') is exactly equivalent to the real core (G) as far as the analyses of Chapter 2 are concerned. For these analyses deal only with the core-edge displacements AA', BB' and do not depend on the shape of the distorted section $A'CB'$. The *equivalent antiplane core* therefore has a shear modulus defined by equation (2.62):

$$G' = \frac{G}{1+\dfrac{E_c}{6E_f}\dfrac{c^2}{t(c+t)}}. \tag{2.62}$$

The procedure is now to use the analyses of Sections 2.2 to 2.8 without modification except that:

(i) EI becomes D (equation 2.2),
(ii) EI_f becomes $E_f I_f$,
(iii) I_f/I becomes $E_f I_f/D$,
(iv) G becomes G' (equation (2.62)).

This procedure yields the correct deflections and face stresses, but the core shear stress which results is τ', the value in the equivalent antiplane core. The maximum shear stress τ_{max} in the real core may be found as follows.

In the real core, from equations (2.59) and (2.60),

$$\frac{\tau_{max}}{BB'} = \frac{2G}{c} \left\{ \frac{1+\dfrac{1}{4}\dfrac{E_c}{E_f}\dfrac{c^2}{td}}{1+\dfrac{1}{6}\dfrac{E_c}{E_f}\dfrac{c^2}{td}} \right\}. \tag{2.63}$$

In the equivalent antiplane core,

$$BB' = \frac{c}{2}\gamma' = \frac{c}{2}\frac{\tau'}{G'}. \tag{2.64}$$

BB' may be eliminated from equations (2.63) and (2.64) to provide the desired value of the maximum shear stress in the real core.

$$\tau_{max} = \tau' \frac{G}{G'} \left\{ \frac{1+\dfrac{1}{4}\dfrac{E_c}{E_f}\dfrac{c^2}{td}}{1+\dfrac{1}{6}\dfrac{E_c}{E_f}\dfrac{c^2}{td}} \right\}. \tag{2.65}$$

If the expression 2.62 for G' is used, this can be simplified:

$$\tau_{max} = \tau'\left(1 + \frac{1}{4}\frac{E_c}{E_f}\frac{c^2}{td}\right). \tag{2.66}$$

These simple modifications to the analyses in Sections 2.2 to 2.8 therefore permit the evaluation of stresses and deflections in sandwich beams with cores which make a substantial contribution to the flexural rigidity of the beam.

2.11. Wide and Narrow Beams

It is convenient to assume in general that sandwich beams bend in a cylindrical manner; that is to say, there is no curvature in the yz-plane (Fig. 2.1). This may well be the case if the load is applied through rigid mountings or if the beam is held down on its supports.

If the beam is narrow in the sense that the width b is less than the core depth c, the lateral expansions and contractions of the faces in the y-direction (associated with the membrane stresses in the x-direction) may take place fairly freely without causing unduly large shear strains in the core in the yz-plane. The faces are therefore mainly in a state of uni-directional stress and the ratio of stress to strain is equal to E. This has been assumed in the analysis of beams in this chapter. The same argument does not apply to the local bending stresses in the faces; each face is a thin plate in cylindrical bending and the ratio of stress to strain is strictly $E/(1-v^2)$. However, these stresses and strains are of secondary importance and it seems reasonable to adopt E throughout in order to avoid undue complication.

If the beam is wide (say $b \gg c$), the lateral expansions and contractions of the faces in the y-direction are severely restricted by the inability of the core to undergo indefinitely large shear deformations in the yz-plane. In this case it may be more reasonable to assume that the strains in the y-direction are zero; this is certainly the only sensible assumption if the plate is infinitely wide. The ratio of stress to strain in the x-direction is therefore $E/(1 - v^2)$ for both the membrane stresses and the local bending stresses, and this value should be used in place of E in all the equations of this chapter.

If anticlastic bending *should* be permitted (that is, if the beam can curve freely in the yz-plane) then E should be used in preference to $E/(1 - v^2)$.

CHAPTER 3

BUCKLING OF SANDWICH STRUTS

3.1. Buckling of Pin-ended Sandwich Strut with Antiplane Core and Thin Faces

The standard analysis of the stability of a uniform axially-loaded pin-ended elastic strut indicates that the strut is unstable when the axial thrust is equal to the Euler load, P_E, where

$$P_E = \frac{\pi^2 D}{L^2}. \qquad (3.1)$$

In this expression D is the flexural rigidity. Strictly, P_E represents the smallest thrust at which the strut will not return to its straight condition after being given some lateral displacement.

In the case of a sandwich strut it is possible for shear deformations to occur in the core. These reduce the stiffness of the sandwich strut and the critical load is correspondingly less than the Euler load defined by equation (3.1). Consider a sandwich with an antiplane core and thin faces, the local bending stiffness of which can be ignored. The flexural rigidity is therefore given by:

$$D = D_1 = E_f b t d^2 / 2. \qquad (3.2)$$

When the strut bends the shear stress distribution is similar to that in Fig. 2.4c.

At the critical value of the axial thrust P there occur two superimposed displacements, w_1 (the ordinary bending displacement) and w_2 (an additional displacement associated with the shear deformation of the core). The buckled strut is shown in Fig. 3.1. At a typical section x the bending moment is:

$$M = P(w_1 + w_2) = -D_1 w_1''. \tag{3.3a}$$

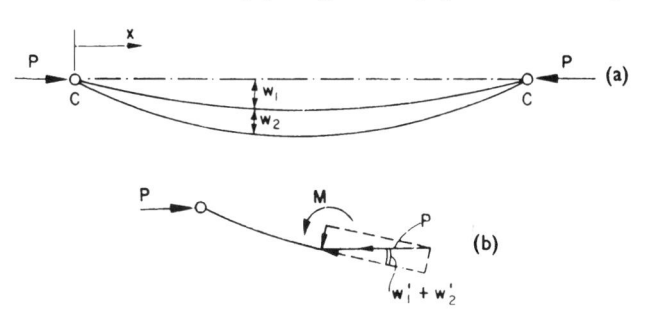

FIG. 3.1. Deflection of an axially-loaded pin-ended sandwich strut.

The horizontal thrust P at the section x has a component $P(w_1' + w_2')$ acting perpendicular to the axis of the strut. This represents the shear force; by equation (2.15a) the shear force is related to the shear deflection w_2:

$$w_2' = \frac{P(w_1' + w_2')}{AG} \qquad (\gamma_0 \text{ is zero}). \tag{3.3b}$$

The term w_2' may be eliminated from equation (3.3a) (differentiated once) and equation (3.3b), to yield a differential equation for w_1:

$$w_1''' + \alpha^2 w_1' = 0, \tag{3.4a}$$

where

$$\alpha^2 = \frac{P}{D_1\{1 - (P/AG)\}}. \tag{3.4b}$$

The solution is of the form

$$w_1 = C_1 \sin \alpha x + C_2 \cos \alpha x + C_3. \tag{3.5a}$$

The total deflection, $w_1 + w_2$, may be obtained from equation (3.3a) by differentiating (3.5a) twice and inserting the result in the right-hand side. Thus:

$$w_1 + w_2 = -\frac{D_1}{P}\{-C_1\alpha^2 \sin \alpha x - C_2\alpha^2 \cos \alpha x\}$$

$$= +\frac{C_1 \sin \alpha x + C_2 \cos \alpha x}{1+(P/AG)}. \tag{3.5b}$$

The boundary condition $w_1 + w_2 = 0$ at $x = 0$ requires C_2 to vanish; the condition $w_1 + w_2 = 0$ at $x = L$ requires that either C_1 vanishes (in which case the strut remains straight) or, if C_1 is non-zero and the strut buckles, then $\sin \alpha L$ vanishes. The function $\sin \alpha L$ vanishes only when $\alpha L = n\pi$ or, from (3.4b),

$$P = \frac{P_E}{1+(P_E/AG)}; \qquad P_E = \frac{\pi^2 D_1}{L^2}. \tag{3.6a}$$

This is the critical load of the sandwich strut; equation (3.6) is essentially the same as the standard result for the critical load of a lattice column. It is often expressed in the equivalent form:

$$\frac{1}{P} = \frac{1}{P_E} + \frac{1}{AG}. \tag{3.6b}$$

When G is finite the critical load is less than the Euler load; when G is infinite the critical load is equal to the Euler load; when G is small the critical load approaches the value AG.

3.2. Buckling of Pin-ended Sandwich Strut with Antiplane Core and Thick Faces

As with a sandwich beam, the faces of a sandwich strut have a certain stiffness in bending about their own separate centroidal axes, which tends to reduce the extent of the secondary displacement w_2 mentioned in the previous section. The true critical load is therefore slightly greater than the value indicated by equation (3.6).

In order to evaluate the true critical load it is convenient to take Section 2.5 as a starting-point because the whole of that section is valid for the bending of both sandwich struts and beams.

Consider a uniform axially-loaded pin-ended elastic sandwich strut which is initially straight and which possesses thick faces and an antiplane core. As in the previous section, when the strut is bent under the action of a thrust P, the total shear force on a typical cross-section is $Q = P(w_1' + w_2')$ (Fig. 3.1b). Substitution in equation (2.27a) yields the following:

$$Q_1'' - a^2 Q_1 = -a^2 P(w_1' + w_2'), \tag{3.7}$$

where a is defined by equation (2.27b).

The slope w_2' is equal to $Q_1/a^2 EI_f$ (equation (2.28)) and Q_1 itself is equal to $-EI w_1'''$ by definition. Equation (3.7) may therefore be converted into a differential equation for w_1:

$$w_1^v - \left(a^2 - \frac{P}{EI_f}\right) w_1''' - \frac{a^2 P}{EI} w_1' = 0. \tag{3.8}$$

If the ends of the faces are free to rotate (e.g. they are not attached to a rigid end-diaphragm) the boundary conditions are $w_1 = w_1'' = 0$ at $x = 0, L$. These conditions are fulfilled by a sinusoidal displacement,

$$w_1 = a_1 \sin \frac{\pi x}{L}. \tag{3.9}$$

Substitution in equation (3.8) and division by $-(\pi/L)\cos(\pi x/L)$ leaves the following:

$$\left\{\frac{\pi^4}{L^4} + \left(a^2 - \frac{P}{EI_f}\right)\frac{\pi^2}{L^2} - \frac{a^2 P}{EI}\right\} a_1 = 0. \tag{3.10}$$

Either $a_1 = 0$, in which case the strut is straight and unbuckled, or the coefficient of a_1 vanishes, which happens when:

$$P = \frac{\dfrac{\pi^4}{L^4} + \dfrac{a^2 \pi^2}{L^2}}{\dfrac{\pi^2}{L^2 EI_f} + \dfrac{a^2}{D}} = \frac{\dfrac{\pi^4}{L^4}(EI)(EI_f)\left(1 - \dfrac{I_f}{I}\right) + \dfrac{\pi^2 EI}{L^2} AG}{\dfrac{\pi^2 EI}{L^2}\left(1 - \dfrac{I_f}{I}\right) + AG}. \tag{3.11}$$

This is the critical load, because non-zero displacements are possible. It may be expressed more simply by writing

$$P_E = \frac{\pi^2 EI}{L^2}, \quad P_{Ef} = \frac{\pi^2 EI_f}{L^2}, \quad P_c = AG = \frac{bd^2}{c}G. \quad (3.12)$$

P_E represents the Euler load of the sandwich strut in the absence of core shear strains; P_{Ef} represents the sum of the Euler loads of the two faces when they buckle as independent struts (i.e. when the core is absent). P_c may be described as the shear buckling load; it is numerically equal to the shear stiffness AG. In terms of these three quantities, the critical load in equation (3.11) takes the following form:

$$P_{cr} = P_E \left\{ \frac{1 + \dfrac{P_{Ef}}{P_c} - \dfrac{P_{Ef}}{P_c} \cdot \dfrac{P_{Ef}}{P_E}}{1 + \dfrac{P_E}{P_c} - \dfrac{P_{Ef}}{P_c}} \right\}. \quad (3.13)$$

The significance of this expression is illustrated in Fig. 3.2 which shows diagrammatically a plot (ABC) of P_{cr} against the strut length L. Also plotted are several curves for extreme cases. For example, DEC represents the Euler load P_E for the strut as a whole; AFG represents the sum (P_{Ef}) of the Euler loads for the two faces, buckling independently; HKC represents equation (3.6) for struts with thin faces.

When the strut is very long, $P_E \to P_{Ef} \to 0$ and equation (3.13) shows that $P_{cr} \to P_E$. This is illustrated by the convergence of curves ABC and DEC at large values of L.

When the strut is very short, $P_E \to \infty$ and equation (3.13) shows that $P_{cr} \to P_{Ef}$. This is illustrated by the convergence of curves AFG and ABC at small values of L.

When the faces are very thin, $P_{Ef} \to 0$ and equation (3.13) coincides with equation (3.6).

When the core is weak in shear, $P_c \to 0$ and $P_{cr} \to P_{Ef}$; that is, the core ceases to provide effective connection between the faces, which buckle as independent struts.

BUCKLING OF SANDWICH STRUTS 53

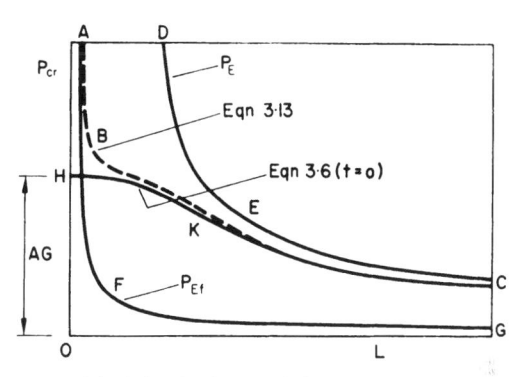

FIG. 3.2. Critical load of an axially-loaded pin-ended sandwich strut.

For some purposes it may be convenient to express the critical load (equation (3.13)) in the alternative form below:

$$\sigma_{\text{cr}} = \frac{\pi^2 E}{12} \left(\frac{c}{L}\right)^2 \left(\frac{t}{c}\right)^2 \left\{ 1 + \frac{3\left(1+\dfrac{c}{t}\right)^2}{\left[1+\dfrac{\pi^2}{2}\dfrac{E}{G}\dfrac{t}{c}\left(\dfrac{c}{L}\right)^2\right]} \right\}. \quad (3.14)$$

In this equation, σ_{cr} is the direct stress in the faces when buckling occurs; it should be noted that equation (3.13) is valid when the faces are of unequal thickness, but equation (3.14) is not.

3.3. Further Consideration of Buckling (Antiplane Core, Thick Faces)

In the previous section it was assumed that the solution of the differential equation for a strut (3.8) could be represented by the simple sine curve, equation (3.9). In general, however, the solution of equation (3.8) takes the following form:

$$w_1 = C_1 \sinh \alpha_1 x + C_2 \cosh \alpha_1 x + C_3 \sin \alpha_2 x + C_4 \cos \alpha_2 x + C_5, \tag{3.15a}$$

where

$$\alpha_1^2 = \tfrac{1}{2}[+B_1 \pm \sqrt{(B_1^2+4B_2)}]; \qquad \alpha_2^2 = \tfrac{1}{2}[-B_1 \pm \sqrt{(B_1^2+4B_2)}] \tag{3.15b}$$

and

$$B_1 = a^2 - \frac{P}{EI_f}; \qquad B_2 = \frac{a^2 P}{EI}. \tag{3.15c}$$

The primary and secondary displacements are related by equations (2.26) and (2.27b):

$$w_2' = -\frac{I}{I_f}\frac{w_1'''}{a^2}.$$

Hence, by differentiation and subsequent integration of equation (3.15a), the secondary deformation can be determined:

$$w_2 = -\frac{I}{a^2 I_f}\{C_1\alpha_1^2 \sinh \alpha_1 x + C_2\alpha_1^2 \cosh \alpha_1 x - C_3\alpha_2^2 \sin \alpha_2 x \\ - C_4\alpha_2^2 \cos \alpha_2 x + C_6\}. \tag{3.16}$$

(If γ_0 is not zero, the bracket contains an additional term $C_7 x$.)

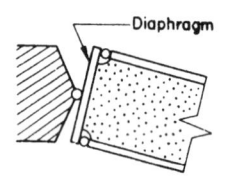

FIG. 3.3. Conditions at a "pinned" end, showing three hinges.

Six boundary conditions are needed for the determination of the constants $C_1 - C_6$. For example, in a pin-ended strut in which the ends of the faces are also free to rotate independently (Fig. 3.3),

(i) $x = 0$, $w_1 = 0$; (ii) $x = 0$, $w_1'' = 0$; (iii) $x = L$, $w_1 = 0$, (iv) $x = L$, $w_1'' = 0$; (v) $x = 0$, $w_2 = 0$; (vi) $x = 0$, $w_2' = 0$.

Of the resulting equations, (ii) and (v) indicate that C_6 is zero. Comparison of (ii) and (vi) leads to the conclusion that C_2 and C_4 are zero, as a consequence of which (i) gives a zero value for C_5. This leaves equations (iii) and (iv) for the determination of C_1 and C_3:

$$C_1 \sinh \alpha_1 L + C_3 \sin \alpha_2 L = 0,$$
$$C_1 \alpha_1^2 \sinh \alpha_1 L - C_3 \alpha_2^2 \sin \alpha_2 L = 0.$$

One solution is $C_1 = C_3 = 0$, which implies that the strut is straight and unbuckled. Another is $C_1 = 0$, $C_3 \neq 0$, $\alpha_2 L = n\pi$, where n is an integer. This last condition gives

$$\alpha_2^2 L^2 = n^2 \pi^2 = \frac{L^2}{2}[-B_1 \pm \sqrt{(B_1^2 + 4B_2)}]. \tag{3.17}$$

The positive sign is appropriate; if the values for B_1 and B_2 are substituted from equation (3.15c), it is possible to show that equation (3.17) reduces to equation (3.11) for the critical load of a pin-ended strut, when $n = 1$.

Because C_3 is the only non-zero constant, equations (3.15a) and (3.16) define the displacements of the buckled strut as follows:

$$w_1 = C_3 \sin \alpha_2 x = C_3 \sin \frac{n\pi x}{L}, \tag{3.18a}$$

$$w_2 = \frac{I}{a^2 I_f} \cdot C_3 \alpha_2^2 \sin \alpha_2 x = \frac{I}{a^2 I_f} C_3 \frac{n^2 \pi^2}{L^2} \sin \frac{n\pi x}{L}. \tag{3.18b}$$

The ratio of the displacements, w_2/w_1, is therefore

$$\frac{I}{a^2 I_f} \cdot \frac{n^2 \pi^2}{L^2},$$

which may also be written

$$\frac{w_2}{w_1} = \frac{n^2 \pi^2}{L^2} \frac{(EI - EI_f)}{AG} = \frac{n^2 \pi^2}{L^2} \left(\frac{Ebtd^2}{2}\right) \bigg/ \left(\frac{Gbd^2}{c}\right)$$
$$= \frac{n^2 \pi^2}{2} \frac{E}{G} \frac{ct}{L^2} = n^2 \xi, \text{ say.} \tag{3.19}$$

Evidently there is a constant relationship between the primary and secondary displacements; this fact is important in the strain energy analyses described in the later chapters. The quantity ξ will also appear frequently. It is a non-dimensional measure of the ratio of the flexural rigidity (neglecting the local bending stiffness of the faces) to the shear stiffness; this is indicated by the middle term in (3.19).

A very similar procedure may be adopted for struts with other end-conditions. For example, in a fixed-ended strut the constants C_1, C_2, C_3 vanish; $C_5 = -C_4$; $C_6 = C_4 \alpha_2^2$; C_4 is non-zero only when $\alpha_2 L = 2n\pi$. It can be shown that the lowest critical load is given by equation 3.13 provided P_E and P_{Ef} are interpreted as $4\pi^2 EI/L^2$ and $4\pi^2 EI_f/L^2$, respectively; P_c remains equal to AG. The deflected form of the fixed-ended strut is as follows:

$$w_1 = -C_4 \left(1 - \cos \frac{2\pi x}{L}\right), \qquad (3.20a)$$

$$w_2 = -4C_4 \xi \left(1 - \cos \frac{2\pi x}{L}\right) \qquad (3.20b)$$

As in the case of the pin-ended strut, the ratio of the primary and secondary displacements is constant for all points along the strut.

3.4. Wrinkling Instability

It is sometimes possible for sandwich struts (or the compression faces of sandwich beams) to fail by a kind of instability associated with short-wavelength ripples in the faces. This *wrinkling* instability may occur at stresses lower than those predicted by the formulae in this chapter. Wrinkling instability is treated separately in Chapter 8.

CHAPTER 4

ANALYSIS OF SANDWICH BEAMS AND STRUTS BY STRAIN ENERGY METHODS

4.1. Introduction, Notation, Assumptions

In this chapter the energy method is applied to the analysis of beams and struts. The analysis serves as a relatively simple introduction to the later use of the method in plate problems, and it also provides an interesting comparison with the results of earlier chapters.

Figure 4.1 shows an elevation of a short length dx of sandwich beam or strut. Each face is of thickness t and of the same material. The core, of thickness c, is antiplane (that is, it contributes nothing to the flexural rigidity of the sandwich and the shear stress is constant over the depth of the core at any particular cross-section). The diagram shows the sandwich before and after displacement w in the direction of the z-axis. $abcde$ represents a line which is normal to the centre-line of the underformed sandwich. During the subsequent displacement, if there were no shear strains, this line would rotate through an angle dw/dx to the position $a'b'c'd'e'$, remaining normal to the centre-line of the sandwich in accordance with the assumptions of the ordinary theory of bending.

If the core only is now allowed to undergo a shear strain γ, the original line moves to a new position $a''b''c'd''e''$, where $d'c'd'' = \gamma$. The lines $a''b''$ and $d''e''$ remain parallel with $a'b'c'd'e'$ because shear strains in the faces are assumed to be negligible.

It is convenient to denote the angle $d''c'z$ by $\lambda(dw/dx)$, where λ

is a coefficient which may have any value between $+1$ and $-t/c$. The value $\lambda = +1$ applies when $\gamma = 0$; the sandwich then bends as a simple composite beam, without shear deformation. The other

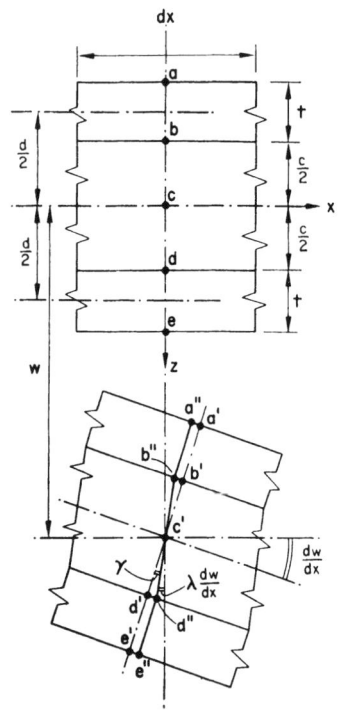

Fig. 4.1. General deformation of short length of sandwich beam.

extreme, $\lambda = -t/c$, is illustrated in Fig. 4.2, which represents a sandwich with a core so flexible in shear as to be incapable of providing any connection between the two faces, which therefore act as independent beams sharing only a common displacement w. The lines ab and de each rotate through an angle dw/dx to new positions $a'b'$, $d'e'$, respectively. The points f and g, which lie on the centre-lines of the faces, have no displacement in the x-direction. Since $dd' = t/2(dw/dx)$, the angle dcd' is equal to $t/c(dw/dx)$. This angle is comparable with the angle $d''c'z = \lambda(dw/dx)$ in Fig. 4.1 except that it is in the opposite sense. Thus $\lambda = -t/c$ in this instance.

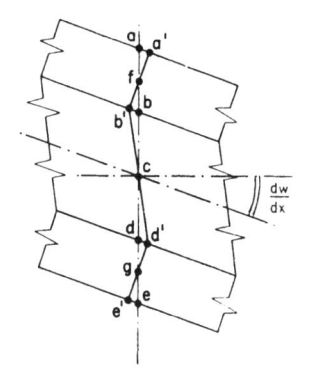

FIG. 4.2. Deformation of short length of sandwich beam with core of negligible shear stiffness.

4.2. Displacements and Strains

The displacements (u) in the x-direction of various points in the distorted cross-section $a''b''d''e''$ (Fig. 4.1) are as follows:

Core ($b''d''$)

$$u = -\lambda z \frac{dw}{dx} \quad -\frac{c}{2} \leq z \leq +\frac{c}{2}. \tag{4.1}$$

Lower face ($d''e''$)

$$\begin{aligned}u &= -\lambda \frac{c}{2}\frac{dw}{dx} - \left(z - \frac{c}{2}\right)\frac{dw}{dx} \\ &= -\left\{\frac{c}{2}(\lambda-1) + z\right\}\frac{dw}{dx} \quad \frac{c}{2} \leq z \leq \frac{h}{2}.\end{aligned} \tag{4.2a}$$

Upper face ($a''b''$)

$$u = -\left\{\frac{c}{2}(1-\lambda) + z\right\}\frac{dw}{dx} \quad -\frac{h}{2} \leq z \leq -\frac{c}{2}. \tag{4.2b}$$

Middle plane of lower face

$$\left(z = \frac{c+t}{2} \text{ in equation (4.2a)}\right)$$

$$u = -\frac{1}{2}(c\lambda+t)\frac{dw}{dx}. \tag{4.3a}$$

Middle plane of upper face

$$\left(z = -\frac{c+t}{2} \text{ in equation (4.2b)}\right)$$

$$u = +\frac{1}{2}(c\lambda+t)\frac{dw}{dx}. \tag{4.3b}$$

All of the z-displacements are equal to w.

Expressions for the strains in the different parts of the sandwich may be obtained by differentiation of the displacements u.

Core strains

The shear strain γ in the core in the xz-plane is given by

$$\gamma = \frac{du}{dz} + \frac{dw}{dx}.$$

Substitution of u from equation (4.1) provides the following:

$$\gamma = (1-\lambda)\frac{dw}{dx}. \tag{4.4}$$

Membrane strain in lower face

This is the strain, e, at the middle-plane of the face; it may be obtained by differentiation of equation (4.3a):

$$e = \frac{du}{dx} = -\frac{1}{2}(c\lambda+t)\frac{d^2w}{dx^2}. \tag{4.5a}$$

Membrane strain in upper face

Similarly, differentiation of equation (4.3b) yields the strain in the middle plane of the upper face.

$$e = +\frac{1}{2}(c\lambda+t)\frac{d^2w}{dx^2}, \tag{4.5b}$$

Local bending strain in lower face

The x-displacement of any point z in the lower face with respect to the middle plane of the lower face is

$$-\left\{\frac{c}{2}(\lambda-1)+z\right\}\frac{dw}{dx}+\frac{1}{2}(c\lambda+t)\frac{dw}{dx} = -\left(z-\frac{c}{2}-\frac{t}{2}\right)\frac{dw}{dx}.$$

The corresponding local bending strain is:

$$e = -\left(z-\frac{c}{2}-\frac{t}{2}\right)\frac{d^2w}{dx^2} \quad \frac{c}{2} \leqslant z \leqslant \frac{h}{2}. \tag{4.6a}$$

(The total direct strain in the lower face is the sum of equations (4.5a) and (4.6a).)

Local bending strain in upper face

Similarly,

$$e = -\left(z+\frac{c}{2}+\frac{t}{2}\right)\frac{d^2w}{dx^2} \quad -\frac{h}{2} \leqslant z \leqslant -\frac{c}{2}. \tag{4.6b}$$

It is important to note that in the derivation of equations (4.5) it has been assumed that λ is independent of x and is therefore constant throughout the length of the sandwich.

4.3 Strain Energy

The strain energy in the faces is defined by the expression

$$U = \frac{E}{2}\int_v e^2 \, dV, \tag{4.7}$$

where E is the modulus of elasticity of the faces, e is the strain in the x-direction, and the integration is carried out over the volume of the faces. The equation implies that the faces are isotropic, that there are no direct stresses in the y- and z-directions and that the shear strain energy in the faces is negligible. The assumption of zero direct stress in the y-direction is permissible only if the sandwich is narrow ($b < c$, say). The assumption that the stress in the z-direction is negligible is usually acceptable because the intensity of transverse distributed loads is much smaller than the magnitude of the stresses in the x-direction to which they give rise. Neglect of the shear strain energy is justified because the faces are in effect solid beams of rectangular cross-section which are shallow in proportion to their span.

The shear strain energy in the core is

$$U = \frac{G}{2} \int_v \gamma^2 \, dV, \qquad (4.8)$$

where G is the shear modulus of the core and γ is the shear strain, both in the xz-plane. The integration is carried out over the volume of the core. By symmetry there are no shear stresses in the xy- and yz-planes; the direct stresses in the x- and y-directions are negligible because the core is assumed to be antiplane. The strain energy due to direct stresses and strains in the z-direction is neglected partly because the transverse load intensity is assumed to be small and partly because the core is assumed to be stiff in the z-direction (as in a honeycomb). These last assumptions might be difficult to justify in the case of a low-density isotropic core, in which significant flattening (squashing) of the core might take place.

In the analysis which follows, suffixes will be avoided wherever possible. It will usually be clear from the context whether a particular quantity such as E or G belongs to the faces or the core.

Shear strain energy in the core

Substitution for γ from equation (4.4) in equation (4.8) yields the following:

$$U = \frac{G}{2} bc \int_0^L (1-\lambda)^2 \left(\frac{dw}{dx}\right)^2 dx. \qquad (4.9)$$

Strain energy due to pure extension of lower or upper face (membrane action)

Substitution for e from equation (4.5a) in equation (4.7) yields the membrane strain energy for the lower face:

$$U = \frac{Ebt}{8} \int_0^L (c\lambda + t)^2 \left(\frac{d^2w}{dx^2}\right)^2 dx. \tag{4.10}$$

An identical expression is obtainable for the upper face from equations (4.5b) and (4.7).

Strain energy due to local bending of lower or upper face

Substitution for e from equation (4.6a) in equation (4.7), and integration over the width and thickness of the face, yield the local bending energy for the lower face:

$$U = \frac{Ebt^3}{12} \int_0^L \left(\frac{d^2w}{dx^2}\right)^2 dx. \tag{4.11}$$

An identical expression is obtainable for the upper face from equations (4.6b) and (4.7).

Potential energy of applied loads

In Fig. 4.3 is shown the centre-line of a sandwich beam in its deflected state under the action of transverse distributed loads

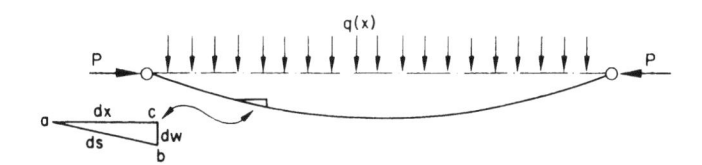

FIG. 4.3. Deflection of a sandwich beam-column.

$q(x)$ and an axial end-load P. The inset diagram shows a short length of centre-line, $ab(=ds)$ and its projection, $ac(=dx)$ on the x-axis. The apparent shortening in the x-direction is $(ds-dx)$. But

ds may be written

$$ds = \sqrt{[dx^2+dw^2]} = dx\sqrt{\left[1+\left(\frac{dw}{dx}\right)^2\right]}.$$

Thus the apparent shortening of a small element of the beam in the x-direction is

$$(ds-dx) = dx\left\{\sqrt{\left[1+\left(\frac{dw}{dx}\right)^2\right]}-1\right\}.$$

Because the slope is small, the square root may be expanded binomially and all but the first two terms may be neglected.

$$(ds-dx) = dx\left\{1+\frac{1}{2}\left(\frac{dw}{dx}\right)^2+ \ldots -1\right\} = \frac{1}{2}\left(\frac{dw}{dx}\right)^2 dx.$$

The apparent shortening of the entire beam in the x-direction is therefore

$$\int_0^L \frac{1}{2}\left(\frac{dw}{dx}\right)^2 dx.$$

As the ends of the beam approach each other by this amount, the end-load P loses potential energy. The change of energy, V_1, may be written

$$V_1 = -\frac{P}{2}\int_0^L \left(\frac{dw}{dx}\right)^2 dx. \qquad (4.12\text{a})$$

This represents the change of potential energy of P due to bending of the member. The end-load P also loses energy as the strut shortens due to axial strain, and there is a corresponding amount of strain energy associated with the axial strain. These quantities do not influence the bending behaviour of the beam and they are omitted from the analysis.

That part of the transverse load $q(x)$ which acts on a short length of beam dx loses potential energy equal to $w \cdot q(x) dx$ as the beam descends through a distance w. The total change of potential energy, V_2, associated with the transverse load is therefore

$$V_2 = -\int_0^L wq(x)\,dx. \tag{4.12b}$$

4.4. Simply-supported Beam-column with Sinusoidal Transverse Load

Consider a simply-supported beam with an axial end-load P and a sinusoidally-distributed transverse load given by

$$q(x) = q_n \sin\frac{n\pi x}{L} \qquad n = 1,2,3,\ldots,\infty. \tag{4.13a}$$

It will be assumed that the transverse displacements w may be written in the form

$$w = a_n \sin\frac{n\pi x}{L} \qquad n = 1,2,3,\ldots,\infty. \tag{4.13b}$$

Equation (4.13b) satisfies the boundary conditions,

$$w = d^2w/dx^2 = 0 \quad \text{at} \quad x = 0, L.$$

The total energy of the system, $U+V$, is represented by the sum of all the expressions (4.9)–(4.12), with (4.10) and (4.11) counted twice. Substitution for w and $q(x)$ from equations (4.13) yields the following:

$$U+V = \int_0^L \left\{ \frac{Gbc}{2}\left[(1-\lambda)a_n\frac{n\pi}{L}\cos\frac{n\pi x}{L}\right]^2 \right.$$
$$+ \frac{Ebt}{4}\left[-(c\lambda+t)a_n\frac{n^2\pi^2}{L^2}\sin\frac{n\pi x}{L}\right]^2 + \frac{Ebt^3}{12}\left[-a_n\frac{n^2\pi^2}{L^2}\sin\frac{n\pi x}{L}\right]^2$$
$$\left. - \frac{P}{2}\left[a_n\frac{n\pi}{L}\cos\frac{n\pi x}{L}\right]^2 - a_nq_n\left[\sin\frac{n\pi x}{L}\right]^2 \right\}dx. \tag{4.14}$$

It is easy to show that

$$\int_0^L \sin^2 \frac{n\pi x}{L} dx = \int_0^L \cos^2 \frac{n\pi x}{L} dx = \frac{L}{2}.$$

With this simplification the expression becomes

$$U+V = a_n^2 \frac{n^2\pi^2}{L^2} \frac{L}{2} \left\{ \frac{Gbc}{2}(1-\lambda)^2 + \frac{Ebt}{4}(c\lambda+t)^2 \frac{n^2\pi^2}{L^2} \right.$$
$$\left. + \frac{Ebt^3}{12} \frac{n^2\pi^2}{L^2} - \frac{P}{2} \right\} - a_n q_n \frac{L}{2}. \qquad (4.15)$$

The total energy $(U+V)$ is expressed here as a function of the unknown amplitude a_n and the corresponding coefficient λ. If the system is to be in equilibrium it is necessary that $(U+V)$ should be stationary with respect to a_n and to λ. Consequently,

$$\frac{\partial}{\partial a_n}(U+V) = \frac{\partial}{\partial \lambda}(U+V) = 0.$$

These two equations may be used to evaluate a_n and λ. For example,

$$\frac{\partial}{\partial \lambda}(U+V) = a_n^2 \frac{n^2\pi^2}{L^2} \frac{L}{2} \left\{ -Gbc(1-\lambda) + \frac{Ebtc}{2}(c\lambda+t)\frac{n^2\pi^2}{L^2} \right\} = 0.$$

(4.16)

Hence

$$\lambda = \frac{1 - \frac{t}{c}\xi n^2}{1 + \xi n^2}, \qquad (4.17a)$$

where

$$\xi = \frac{\pi^2}{L^2} \frac{E}{G} \frac{ct}{2}. \qquad (4.17b)$$

The dimensionless coefficient ξ is the product of π^2/L^2 and the ratio of the flexural rigidity $Ebtd^2/2$ (neglecting the local bending stiffness of the faces) to the shear stiffness, Gbd^2/c. This value for λ may be inserted in equation (4.15) and, as a result, the condition for $(U+V)$ to be stationary with respect to a_n is as follows:

$$\frac{\partial}{\partial a_n}(U+V) = 2a_n \frac{n^2\pi^2}{L^2} \frac{L}{2} \left\{ \frac{Gbd^2}{2c} \frac{n^2\xi}{1+n^2\xi} + \frac{Ebt^3}{12} \frac{n^2\pi^2}{L^2} - \frac{P}{2} \right\}$$
$$- q_n \frac{L}{2} = 0.$$

Or $\quad a_n \dfrac{n^2\pi^2}{L^2} \left\{ \dfrac{n^2\pi^2}{L^2} \dfrac{Ebd^2t}{2} \left[\dfrac{1}{1+n^2\xi} + \dfrac{t^2}{3d^2} \right] - P \right\} = q_n.$ \quad (4.18)

This equation defines the amplitude a_n of the transverse displacements.

When an axial thrust P is applied in the absence of a transverse load, both sides of equation (4.18) must vanish. Either a_n is zero and the strut is straight and unbuckled, or P must take a value which causes the coefficient of a_n to vanish. In the latter case P is the critical load (P_n) at which the strut buckles and at which non-zero values of a_n are possible. If the coefficient of a_n is to vanish P_n must have the value:

$$P_n = \frac{n^2\pi^2}{L^2} \frac{Ebd^2t}{2} \frac{1}{\chi}, \qquad (4.19a)$$

where

$$\frac{1}{\chi} = \left\{ \frac{1}{1+n^2\xi} + \frac{t^2}{3d^2} \right\}. \qquad (4.19b)$$

Equation (4.19a) may also be written in a form of which equation (3.13) represents the special case $n = 1$:

$$P_n = n^2 P_E \left\{ \frac{1 + \dfrac{n^2 P_{Ef}}{P_c} - \dfrac{n^2 P_{Ef}}{P_c} \dfrac{P_{Ef}}{P_E}}{1 + \dfrac{n^2 P_E}{P_c} - \dfrac{n^2 P_{Ef}}{P_c}} \right\}, \qquad (4.19c)$$

where

$$P_E = \frac{\pi^2 EI}{L^2}, \quad P_{Ef} = \frac{\pi^2 EI_f}{L^2}, \quad P_c = \frac{Gbd^2}{c},$$
$$I = \frac{bd^2t}{2} + \frac{bt^3}{6}, \quad I_f = \frac{bt^3}{6}.$$

When a sinusoidal transverse load is applied in the absence of end-load, equation (4.18) defines the amplitude of the resulting sinusoidal displacement as:

$$a_n = q_n \frac{L^4}{n^4\pi^4} \frac{2}{Eb\,d^2t} \chi. \tag{4.20}$$

When the beam is subjected simultaneously to transverse and end loads, a_n must be found from equation (4.18) in full. It is sometimes convenient to write the equation in an alternative manner:

$$a_n = q_n \frac{L^4}{n^4\pi^4} \frac{2}{Eb\,d^2t} \chi \left(1 - \frac{P}{P_n}\right)^{-1}. \tag{4.21}$$

In comparing this equation with equation (4.20) it may be observed that the amplitude of the deformation is amplified by a factor $(1 - P/P_n)^{-1}$ in the presence of an end-load P. As might be expected, the amplitude tends to infinity as the end-load approaches the critical value P_n.

Equations (4.19a) and (4.20) reduce to familiar and expected forms in certain extreme conditions, which are listed in Table 4.1.

This method of minimizing the total energy with respect to a parameter such as λ was first used for sandwich struts by Williams, Leggett and Hopkins.[8.1]

4.5. Evaluation of Stresses due to Combined End-load and Sinusoidal Transverse Load

In the previous section a simply supported beam was subjected to an axial load P and a sinusoidally distributed transverse load $q_n \sin n\pi x/L$. The amplitude a_n of the corresponding sinusoidal transverse displacements was determined (equations (4.20), (4.21)). With this information available it is a simple matter to evaluate the stresses in the faces and in the core.

TABLE 4.1. CRITICAL LOADS, AMPLITUDES AND STRESSES IN THREE EXTREME CASES

	Thin faces $\dfrac{t}{d} \to 0 \quad \xi \to \dfrac{\pi^2 EI}{L^2 G\, db}$ [a]	Core flexible in shear $G \to 0 \quad \xi \to \infty$	Core rigid in shear $G \to \infty \quad \xi \to 0$	
nth critical load, P_n [c] (equation (4.19a))	$\dfrac{n^2 P_B}{1 + n^2 P_B / P_c}$ [b, e]	$\dfrac{n^2 \pi^2 EI_f}{L^2}$	$\dfrac{n^2 \pi^2 EI}{L^2}$ [e]	
Amplitude a_n [d] (equation (4.20))	$q_n \left(\dfrac{L^4}{n^4 \pi^4 EI} + \dfrac{L^2}{n^2 \pi^2 Gbd} \right)$	$\dfrac{q_n L^4}{n^4 \pi^4 EI_f}$ [e]	$\dfrac{q_n L^4}{n^4 \pi^4 EI}$ [e]	
Membrane stress in lower face (equation (4.25)) [d]	1	0	$1 - \dfrac{I_f}{I}$	$\times \dfrac{q_n}{bdt} \dfrac{L^2}{n^2 \pi^2} \sin \dfrac{n\pi x}{L}$ [e]
Local bending stress at lower surface of lower face (equation (4.26)) [d]	indeterminate	1	$\dfrac{I_f}{I}$	$\times \dfrac{q_n}{(bt^2/3)} \dfrac{L^2}{n^2 \pi^2} \sin \dfrac{n\pi x}{L}$ [e]
Core shear stress (equation (4.27)) [d]	1	0	$1 - \dfrac{I_f}{I}$	$\times \dfrac{q_n}{bd} \dfrac{L}{n\pi} \cos \dfrac{n\pi x}{L}$ [e]

[a] Compare equation (3.6a) ($n = 1$).
[b] The first and second components represent bending and shear displacements respectively.
[c] In the absence of transverse load.
[d] In the absence of end-load.
[e] q_n is the amplitude of the sinusoidal transverse load in the nth mode.

The membrane strain in the lower face is given by equation (4.5a). Multiplication by E, and substitution for w from equation (4.13b) gives the membrane stress in the lower face:

$$\sigma = +\frac{E}{2}(c\lambda+t)\,a_n \cdot \frac{n^2\pi^2}{L^2}\sin\frac{n\pi x}{L}. \tag{4.22}$$

This does not include the direct stress due to end-load.

The local bending strain at the lower surface of the lower face is obtained by writing $z = c/2 + t$ in equation (4.6a). Multiplication by E, and substitution for w from equation (4.13b), gives the corresponding local bending stress:

$$\sigma = +\frac{Et}{2}a_n\frac{n^2\pi^2}{L^2}\sin\frac{n\pi x}{L}. \tag{4.23}$$

The shear strain in the core is given by equation (4.4). Multiplication by G, and substitution for w from equation (4.13b) gives the corresponding core shear stress:

$$\tau = G(1-\lambda)\,a_n\frac{n\pi}{L}\cos\frac{n\pi x}{L}. \tag{4.24}$$

Equations (4.22)–(4.24) may be written in a more convenient form by substitution of λ from equation (4.17a) and a_n from equation (4.20) (supposing, for the moment, that $P = 0$).

Membrane stress in lower face

$$\sigma = \frac{q_n}{bdt}\frac{L^2}{n^2\pi^2}\frac{\chi}{1+\xi n^2}\sin\frac{n\pi x}{L}. \tag{4.25}$$

Local bending stress at lower surface of lower face

$$\sigma = \frac{q_n}{(bt^2/3)}\frac{L^2}{n^2\pi^2}\chi\sin\frac{n\pi x}{L}. \tag{4.26}$$

Core shear stress

$$\tau = \frac{q_n}{bd} \frac{L}{n\pi} \frac{\chi}{1+\xi n^2} \cos \frac{n\pi x}{L}. \quad (4.27)$$

If the end-load is not zero, these stresses must all be factored by $(1-P/P_n)^{-1}$. The direct stresses σ are tensile when positive.

The limiting forms of these equations in the three extreme cases are shown in Table 4.1. It should be noted that when the faces are thin the membrane stresses in the faces and the core shear stresses are independent of the elastic properties of the face and core. This result was foreshadowed in Chapter 2 where, for sandwiches with thin faces, stresses were calculated directly from the bending moments and shear forces derived by elementary statics.

4.6. Deflections and Stresses due to Non-sinusoidal Transverse Loads

The analysis of the previous section may be extended readily to permit the determination of deflections and stresses due to transverse loads which are not sinusoidally distributed. For simplicity, axial loads will not be considered in the first instance.

Suppose that a simply-supported sandwich beam supports a transverse load of varying intensity $q(x)$. This function may be expressed as a Fourier series:

$$q(x) = \sum_{n=1}^{\infty} q_n \sin \frac{n\pi x}{L}, \quad (4.28)$$

where

$$q_n = \frac{2}{L} \int_0^L q(x) \sin \frac{n\pi x}{L} dx. \quad (4.29)$$

The Fourier coefficients are therefore known in any particular problem. A typical load component, $q_n \sin n\pi x/L$, of the series

(4.28) causes a transverse deflection of the form $a_n \sin n\pi x/L$ (equation (4.20)).[†] The principle of superposition is valid with respect to transverse loads and their effects. Consequently the deflection due to the given transverse load $q(x)$ is equal to the sum of the deflections due to each of the load components $q_n \sin n\pi x/L$ for $n = 1, 2, 3, \ldots, \infty$. Similar considerations apply to the stresses.

Consider, for example, the load case which is likely to be of greatest practical interest; that is, where the load is uniformly distributed with an intensity q. It is easy to show that in this instance

$$q = \frac{4q}{n\pi} \qquad n = 1, 3, 5, \ldots, \infty$$
$$= 0 \qquad n = 2, 4, 6, \ldots, \infty. \qquad (4.30)$$

Insertion of this value in equations (4.13b) and (4.20) leads to the following equation for deflection:

$$w = \frac{8qL^4}{Ebd^2 t\pi^5} \sum \frac{\chi}{n^5} \sin \frac{n\pi x}{L} dx \qquad n = 1, 3, 5, \ldots, \infty. \qquad (4.31)$$

Similarly, stresses may be expressed as the sums of expressions (4.25)–(4.27) in turn.

Generally, only the peak values of deflections and stresses are of interest. All of the components of the series for deflections and face stresses (4.31, 4.25, 4.26) contain the term $\sin n\pi x/L$, the

[†] This simple result is essentially due to the orthogonality of the modes chosen; that is, because

$$\int_0^L \sin \frac{n\pi x}{L} \sin \frac{m\pi x}{L} dx = \int_0^L \cos \frac{n\pi x}{L} \cos \frac{m\pi x}{L} dx = 0, \quad \text{when} \quad n \neq m.$$

A more precise indication of the importance of orthogonal modes is given in Sections 5.4 and 6.5 in relation to plates.

peak value of which is $(-1)^{(n-1)/2}$ at $x = L/2$ (n takes odd values only). All the components of the series for the core shear stress (4.27) contain the term $\cos n\pi x/L$, the peak value of which is $+1$ at $x = 0$.

Consequently, from equations (4.30), (4.31) and (4.25)–(4.27) the peak deflections and stresses due to a uniformly distributed transverse load q are as follows:

Deflection:

$$w_{\max} = \frac{4qL^4}{\pi^5 D_1} \sum \frac{\chi}{n^5} (-1)^{(n-1)/2} \quad \text{where} \quad D_1 = \frac{Ebtd^2}{2}. \tag{4.32}$$

Membrane stress in face:

$$\sigma_{\max} = \frac{4q}{bdt} \frac{L^2}{\pi^3} \sum \frac{\chi}{n^3(1+n^2\xi)} (-1)^{(n-1)/2}. \tag{4.33}$$

Local bending stress in face:

$$\sigma_{\max} = \frac{4q}{(bt^2/3)} \frac{L^2}{\pi^3} \cdot \frac{t^2}{3d^2} \sum \frac{\chi}{n^3} (-1)^{(n-1)/2}. \tag{4.34}$$

Shear stress in core:

$$\tau_{\max} = \frac{4q}{bd} \frac{L}{\pi^2} \sum \frac{\chi}{n^2(1+n^2\xi)}. \tag{4.35}$$

All of the summations extend over the values $n = 1, 3, 5, \ldots, \infty$. The effect of axial thrust P is to amplify the nth term in each series by a factor $(1-P/P_n)^{-1}$ and to introduce an axial compressive stress $P/2bt$.

A similar procedure can be adopted for other types of loading. However, it may not always be so easy to determine the positions of maximum deflection, bending moment and shear force. Furthermore, when the load contains asymmetrical Fourier components (n even) the value of γ_0, the overall core shear strain discussed in Section 2.4, may not always be zero.

When the faces are thin and $t \ll d$, then $\chi \to (1+\xi n^2)$. Equations (4.32), (4.33) and (4.35) then reduce to the following form:

$$w_{max} = \frac{4qL^5}{\pi^5 D_1} \sum \frac{(-1)^{(n-1)/2}}{n^5} + \frac{4qL^2}{\pi^3 AG} \sum \frac{(-1)^{(n-1)/2}}{n^3}, \quad (4.36)$$

$$\sigma_{max}(\text{membrane}) = \frac{4qL^2}{bdt\pi^3} \sum \frac{(-1)^{(n-1)/2}}{n^3}, \quad (4.37)$$

$$\tau_{max} = \frac{4qL}{\pi^2 bd} \sum \frac{1}{n^2}. \quad (4.38)$$

The local bending stress in the face (4.34) is indeterminate when $t \to 0$. The first and second terms in the expression for w_{max} represent respectively the bending deflection $5qL^4/384D_1$ and the shear deflection $(qL^2/8AG)$. The values of the maximum face and core stresses, from elementary beam analysis, are $qL^2/8bdt$ and $qL/2bd$ respectively.

The number of terms which must be evaluated in the series in equations (4.36)–(4.38) in order to achieve satisfactory approximations to these known results can be discovered from Table 4.2.

TABLE 4.2. PERCENTAGE ERRORS DUE TO NEGLECT OF HIGHER TERMS IN SERIES IN EQUATIONS (4.36)–(4.38)

Sum to $n=$	$\sum \dfrac{(-1)^{(n-1)/2}}{n^5}$	$\sum \dfrac{(-1)^{(n-1)/2}}{n^3}$	$\sum n^{\frac{1}{2}}$
1	+0·383	+3·2	−18·9
3	−0·029	−0·6	− 9·9
5	+0·003	+0·2	− 6·7
7	—	−0·1	− 5·0
9	—	—	− 4·0
11	—	—	− 3·4
13	—	—	− 2·9
15	—	—	− 2·5

For example, the bending deflection (first term on right-hand side of (4.36)) obtained by taking only the first term of the appropriate series is a mere 0.383% high. Similarly, the shear deflection (second term on right-hand side of (4.36)) and the membrane stress in the face (4.37) obtained by taking only the first term of the series

$$\sum \frac{(-1)^{(n-1)/2}}{n^3}$$

are 3.2% too high; this is probably accurate enough for most practical purposes. On the other hand, the series for the shear stress (4.38) converges quite slowly. It is necessary to add the first three terms ($n = 1, 3, 5$) before the error reaches a level which might be acceptable (-6.7%); a really accurate approximation involves summing the series as far as $n = 15$ or further.

CHAPTER 5

BENDING AND BUCKLING OF ISOTROPIC SANDWICH PANELS WITH VERY THIN IDENTICAL FACES (RITZ METHOD)

5.1. Introduction

In the earlier chapters attention has been confined to beams and struts which bend in one direction only. This chapter is concerned instead with plates supported on four sides, bending in two directions simultaneously. In some cases the plate may be subjected to loads normal to the plane of the plate. The problem is then to determine the normal displacements and the stresses in the plate. In other cases the plate may support loads applied at the edges, in the plane of the plate. The plate will remain flat until the edge forces reach a certain critical magnitude, at which point the plate is in neutral equilibrium and is capable of sustaining a buckled shape. The problem in this case is to determine the critical values of the edge loads at which buckling becomes possible. The two problems will be referred to as the bending and buckling problems respectively. Much of the analysis is common to both problems and it will therefore be convenient to develop equations which can be used to represent the solution to either problem or, indeed, to the problem of combined bending and buckling.

In this chapter the analysis is restricted to sandwich plates with very thin faces. This has been done in order to present a simple solution with a minimum of algebra which will nevertheless demonstrate the essential character of the behaviour of sandwich panels. The effects of thick faces will be treated in Chapter 6.

The middle plane of the plate, of dimensions $a \times b$, lies in the xy-plane (Fig. 5.1). The faces are each of thickness t and the core is of thickness c. The following assumptions are made.

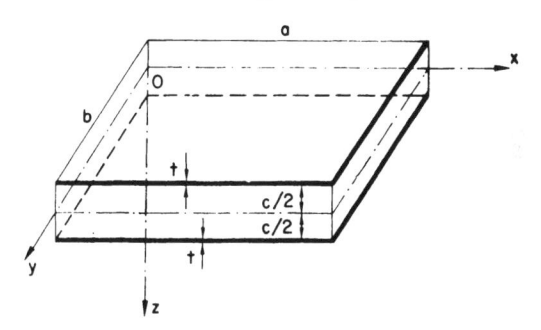

FIG. 5.1. Dimensions of sandwich panel with thin equal faces.

1. Stresses in the faces and the core in the z-direction are of no importance and may be neglected.
2. In the xy-plane the faces and the core are isotropic.
3. In the xy-plane the core is much less stiff than the faces. The contribution of the core to the flexural rigidity of the sandwich may therefore be neglected and the core shear stresses at any position (x, y) are therefore independent of z and constant throughout the depth of the core (see Section 2.1).
4. Deflections are small. Therefore, not only is the ordinary theory of bending valid, but also there is no stretching of the middle plane of the plate when transverse displacements occur. This implies that the forces N_x, N_y, N_{xy} in the plane of the plate are not altered by displacements in the z-direction.
5. The faces are thin in comparison with the thickness of the core. Consequently the local bending stiffness of the faces

is negligible, and it is *also* permissible to write $c = d$ (Fig. 2.1).

The sign convention which will be adopted for plates in this and later chapters is shown in Fig. 5.2, which shows the positive senses of bending and twisting moments (M_x, M_y, M_{xy}, M_{yx}), shear forces (Q_x, Q_y) and membrane forces (N_x, N_y, N_{xy}, N_{yx}).

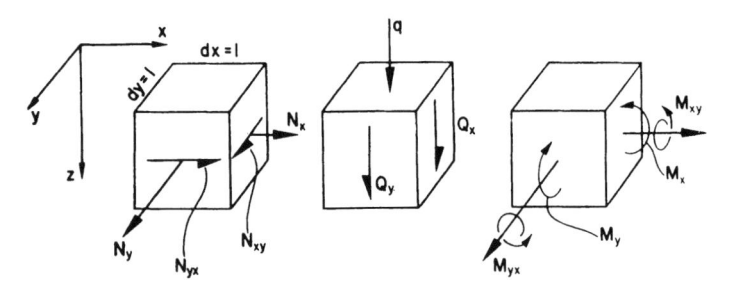

Fig. 5.2. Sign conventions for sandwich panel; positive senses shown.

Each of these quantities refers to a unit length; for example, the shear forces on an element of dimensions $dxdy$ are $Q_x dy$ and $Q_y dx$. This convention coincides with that used by Timoshenko[35.3, 35.14]; it is also compatible with the convention used earlier in the book for beams which bend in the zx-plane. Shear stresses and strains will be considered as positive when they operate in the positive senses of Q_x and Q_y.

5.2. Displacements and Strains

The procedure for the evaluation of the displacements and strains in the core and in the faces is similar to that used in Chapter 4, except that some simplification is achieved here because the faces are very thin. In Fig. 5.3 is shown a section of the deflected plate, parallel with the zx-plane, after undergoing a displacement in the z-direction. A typical point in the deformed core is represented by F, a distance z below the centre, A. The centre line AG and the normal AE have both rotated through an angle $\partial w/\partial x$, but

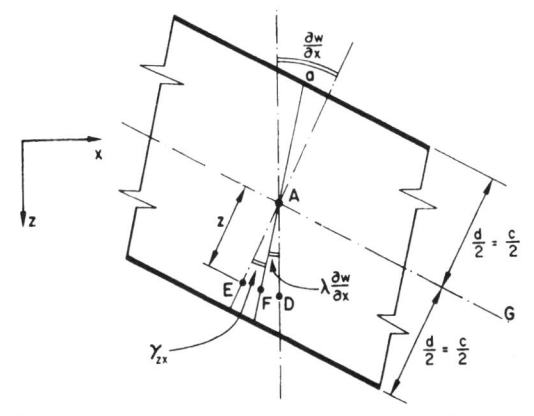

FIG. 5.3. Section through deflected sandwich panel in zx-plane.

the line AF has rotated through a smaller angle, $\lambda \, \partial w/\partial x$. Consequently the shear strain γ_{zx}, which is the angle EAF, is given by

$$\gamma_{zx} = (1-\lambda)\frac{\partial w}{\partial x}. \tag{5.1}$$

The quantity λ may take any value between $+1$ (core rigid in shear) and zero (core completely flexible), depending on the properties and proportions of the sandwich.[†]

The possibility of non-zero values of γ_0 will not be considered in this chapter.

Because the rotation of AF is $\lambda \, \partial w/\partial x$,[‡] the displacement of F in the x-direction is given by

$$u = -z\lambda \frac{\partial w}{\partial x}. \tag{5.2}$$

[†] Had the faces not been very thin, the limits for λ would have been $+1$ and $-t/c$; see Section 4.1.
[‡] This is another possible way of defining λ.

80 ANALYSIS OF STRUCTURAL SANDWICH PANELS

In a similar way the shear strain γ_{yz} in the yz-plane may be written

$$\gamma_{yz} = (1-\mu)\frac{\partial w}{\partial y}. \tag{5.3}$$

The displacement of a point such as F in the y-direction is

$$v = -z\mu\frac{\partial w}{\partial y}. \tag{5.4}$$

The direct strains in the x- and y- directions (tension positive) are

$$e_x = \frac{\partial u}{\partial x} = -z\lambda\frac{\partial^2 w}{\partial x^2}; \quad e_y = \frac{\partial v}{\partial y} = -z\mu\frac{\partial^2 w}{\partial y^2}. \tag{5.5}$$

The shear strain in the xy-plane is

$$\gamma_{xy} = \frac{\partial u}{\partial y} + \frac{\partial v}{\partial x} = -z(\lambda+\mu)\frac{\partial^2 w}{\partial x\,\partial y}. \tag{5.6}$$

The displacements and strains in the faces are obtained by writing $z = \mp d/2$ in equations (5.2)–(5.6); in general it will be convenient to let the upper and lower signs refer to the upper and lower faces respectively. It should be noted that λ and μ have been treated as being independent of x and y during differentiation.

5.3. Strain Energy

The strain energy of an isotropic solid in which $\sigma_z = 0$ is given by

$$U = \frac{E}{2g}\int_v (e_x^2+e_y^2+2\nu e_x e_y)\,dV + \frac{G}{2}\int_v (\gamma_{xy}^2+\gamma_{yz}^2+\gamma_{zx}^2)\,dV, \tag{5.7}$$

where the integration is carried out over the volume of the solid, E, G, ν are the modulus of elasticity, shear modulus and Poisson's ratio for the material and $g = 1 - v^2$. The significance of equation (5.7) is discussed in Appendix I.

Strain energy of core (U_c)

The core is assumed to have no direct stiffness in the x- and y-directions, so the strain energy associated with e_x and e_y in equation (5.7) is zero. The core is also assumed to have no shear stiffness in the xy-plane, so the strain energy associated with γ_{xy} is also zero. This leaves the terms containing γ_{yz} and γ_{zx} only, the values of which are defined by equations (5.1) and (5.3).

$$U_c = \frac{G}{2} \int_{-\frac{d}{2}}^{+\frac{d}{2}} \int_0^a \int_0^b \left\{ (1-\mu)^2 \left(\frac{\partial w}{\partial y}\right)^2 + (1-\lambda)^2 \left(\frac{\partial w}{\partial x}\right)^2 \right\} dy\, dx\, dz$$

$$= \frac{Gd}{2} \int_0^a \int_0^b \left\{ (1-\mu)^2 \left(\frac{\partial w}{\partial y}\right)^2 + (1-\lambda)^2 \left(\frac{\partial w}{\partial x}\right)^2 \right\} dy\, dx. \qquad (5.8)$$

Strain energy of faces (U_f)

The faces are assumed to be perfectly rigid in shear in the yz- and zx-planes, so the terms containing γ_{yz} and γ_{zx} in equation (5.7) vanish. The strain energy of the lower face may be obtained by inserting in equation (5.7) the values of e_x, e_y, γ_{xy} from equations (5.5) and (5.6), writing $z = +d/2$ at the same time:

S.E. of lower face =

$$= \frac{E}{2g} \int_v \left\{ \frac{d^2}{4} \lambda^2 \left(\frac{\partial^2 w}{\partial x^2}\right)^2 + \frac{d^2}{4} \mu^2 \left(\frac{\partial^2 w}{\partial y^2}\right)^2 + 2v \frac{d^2}{4} \lambda\mu \left(\frac{\partial^2 w}{\partial x^2} \cdot \frac{\partial^2 w}{\partial y^2}\right) \right\} dV$$

$$+ \frac{G}{2} \int_v \frac{d^2}{4} (\lambda+\mu)^2 \left(\frac{\partial^2 w}{\partial x\, \partial y}\right)^2 dV.$$

The strain energy of the upper face is the same. The total strain energy of both faces (U_f) is therefore obtained by integrating over the thickness t and doubling the expression given above. It will also be convenient to write $G = E/2(1+\nu)$.

$$U_f = \frac{Ed^2t}{4g} \int_0^a \int_0^b \left\{ \lambda^2 \left(\frac{\partial^2 w}{\partial x^2}\right)^2 + \mu^2 \left(\frac{\partial^2 w}{\partial y^2}\right)^2 + 2\nu\lambda\mu \, \frac{\partial^2 w}{\partial x^2} \cdot \frac{\partial^2 w}{\partial y^2} \right.$$
$$\left. + \left(\frac{1-\nu}{2}\right)(\lambda+\mu)^2 \left(\frac{\partial^2 w}{\partial x \, \partial y}\right)^2 \right\} dy \cdot dx. \tag{5.9}$$

It is important to recognize that G, in equation (5.8), refers to the core material; E and ν in equation (5.9) refer to the face material.

Potential energy of applied loads (N_x, q)

It was shown in Section 4.3 that when a beam of length L is subjected to transverse deformations w, the ends of the beam approach each other by an amount

$$\frac{1}{2} \int_0^L \left(\frac{dw}{dx}\right)^2 dx.$$

Consider now a narrow strip of the plate in Fig. 5.1, parallel with the x-axis, and of width dy. In the same way, the ends of this strip approach each other by an amount

$$\frac{1}{2} \int_0^a \left(\frac{\partial w}{\partial x}\right)^2 dx$$

as the plate bends.

If a tensile force N_x per unit length is applied to the edges $x = 0$, $x = a$ in the plane of the plate, the force applied to the strip is $N_x \, dy$ and the change of potential energy of this force, as the plate bends, is equal to

$$+ \frac{N_x \, dy}{2} \int_0^a \left(\frac{\partial w}{\partial x}\right)^2 dx.$$

BENDING AND BUCKLING OF ISOTROPIC SANDWICH PANELS 83

The total change of potential energy V_1 of the edge forces N_x is obtained by integrating over the width of the plate, from $y = 0$ to $y = b$.

$$V_1 = +\frac{N_x}{2} \int_0^a \int_0^b \left(\frac{\partial w}{\partial x}\right)^2 dy\, dx. \tag{5.10}$$

If the plate also supports a uniform transverse pressure q in the z-direction, then the change of potential energy, V_2, associated with q as the plate bends is

$$V_2 = -\int_0^a \int_0^b wq\, dy\, dx. \tag{5.11}$$

5.4. Plate with Isotropic Faces and Core

It will be assumed that the displacements w for a simply-supported rectangular plate may be expressed in the form

$$w = \sum_{m=1}^{\infty} \sum_{n=1}^{\infty} a_{mn} \sin\frac{m\pi x}{a} \sin\frac{n\pi y}{b}, \tag{5.12}$$

where a_{mn} is the unknown amplitude of the m, nth mode of deformation. Equation (5.12) satisfies the boundary conditions for a simply-supported plate; the importance of the boundary conditions is discussed in Section 6.5.

The total energy of the system, $U+V$, may be obtained by adding the expressions (5.8)–(5.11) and substituting the series (5.12) for w. Consider, for example, the contribution of the first of the two terms which constitute U_c (equation (5.8)):

$$U_{c1} = \frac{Gd}{2} \int_0^a \int_0^b \left\{(1-\mu)\left(\frac{\partial w}{\partial y}\right)\right\}^2 dy\, dx. \tag{5.13}$$

Substitution for w from equation (5.12) gives

$$U_{c1} = \frac{Gd}{2} \int_0^a \int_0^b \left\{ \sum_{m=1}^{\infty} \sum_{n=1}^{\infty} (1 - \mu_{mn}) \, a_{mn} \frac{n\pi}{b} \cdot \sin \frac{m\pi x}{a} \cos \frac{n\pi y}{b} \right\}^2 dy \, dx. \tag{5.14}$$

It is important to note that there is a distinct value of $\mu(=\mu_{mn})$ for each mode of deformation. When the series is squared it will be found that the integrals of the cross-product terms vanish, because of the orthogonal properties of the chosen sinusoidal functions. This leaves only the squared terms:

$$U_{c1} = \frac{Gb}{2} \int_0^a \int_0^b \left\{ \sum_{m=1}^{\infty} \sum_{n=1}^{\infty} (1 - \mu_{mn})^2 \, a_{mn}^2 \frac{n^2 \pi^2}{b^2} \sin^2 \frac{m\pi x}{a} \cos^2 \frac{n\pi y}{b} \right\} dy \, dx. \tag{5.15}$$

This series may be integrated term by term. The integral

$$\int_0^a \int_0^b \sin^2 \frac{m\pi x}{a} \cos^2 \frac{n\pi y}{b} \, dy \, dx$$

is equal to $ab/4$ for all values of m and n. Hence

$$U_{c1} = \frac{Gd}{2} \sum_{m=1}^{\infty} \sum_{n=1}^{\infty} (1 - \mu_{mn})^2 \, a_{mn}^2 \frac{n^2 \pi^2}{b^2} \cdot \frac{ab}{4}. \tag{5.16}$$

A similar process may be applied to each of the energy terms U_c, U_f, V_1 and V_2. For this purpose it is useful to take note of the following relationships, which are derived from equation (5.12):

$$\int_0^a \int_0^b \left(\frac{\partial^2 w}{\partial x^2} \right)^2 dy \, dx = \sum \sum a_{mn}^2 \frac{m^4 \pi^4}{a^4} \frac{ab}{4}, \tag{5.17a}$$

$$\int_0^a \int_0^b \left(\frac{\partial^2 w}{\partial y^2} \right)^2 dy \, dx = \sum \sum a_{mn}^2 \frac{n^4 \pi^4}{b^4} \frac{ab}{4}, \tag{5.17b}$$

$$\int_0^a \int_0^b \left(\frac{\partial^2 w}{\partial x^2}\right)\left(\frac{\partial^2 w}{\partial y^2}\right) dy\, dx = \sum \sum a_{mn}^2 \frac{m^2 n^2 \pi^4}{a^2 b^2} \frac{ab}{4}, \quad (5.17c)$$

$$\int_0^a \int_0^b \left(\frac{\partial^2 w}{\partial x\, \partial y}\right)^2 dy\, dx = \sum \sum a_{mn}^2 \frac{m^2 n^2 \pi^4}{a^2 b^2} \frac{ab}{4}, \quad (5.17d)$$

$$\int_0^a \int_0^b \left(\frac{\partial w}{\partial x}\right)^2 dy\, dx = \sum \sum a_{mn}^2 \frac{m^2 \pi^2}{a^2} \frac{ab}{4}, \quad (5.17e)$$

$$\int_0^a \int_0^b \left(\frac{\partial w}{\partial y}\right)^2 dy\, dx = \sum \sum a_{mn}^2 \frac{n^2 \pi^2}{b^2} \frac{ab}{4}, \quad (5.17f)$$

$$\int_0^a \int_0^b w\, dy\, dx = \sum \sum \frac{4 a_{mn}}{\pi^2} \frac{ab}{mn} \quad (m, n \text{ both odd}) \quad (5.17g)$$

$$= 0 \text{ (otherwise)}.$$

Substitution of these values in equations (5.8)–(5.11) in the manner illustrated above yields equations (5.18) for the various energy terms U_c, U_f, V_1, V_2.

For simplicity only the (m, n)th terms are shown below; also the suffixes m, n are omitted from λ and μ, it being understood that λ and μ take different values for the different modes m, n.

$$(U_c)_{mn} = GA_1 \left\{ (1-\lambda)^2 \frac{m^2}{a^2} + (1-\mu)^2 \frac{n^2}{b^2} \right\} a_{mn}^2, \quad (5.18a)$$

$$(U_f)_{mn} = EA_2 \left\{ \lambda^2 \frac{m^4}{a^4} + \mu^2 \frac{n^4}{b^4} + 2\nu\lambda\mu \frac{m^2 n^2}{a^2 b^2} + \frac{1-\nu}{2}(\lambda+\mu)^2 \frac{m^2 n^2}{a^2 b^2} \right\} a_m^2, \quad (5.18b)$$

$$(V_1)_{mn} = \frac{N_x}{2} \pi^2 a_{mn}^2 \frac{ab}{4} \frac{m^2}{a^2}, \quad (5.18c)$$

$$(V_2)_{mn} = -4q\frac{a_{mn}}{\pi^2}\frac{ab}{mn} \quad (m, n \text{ odd}), \tag{5.18d}$$

$$A_1 = \frac{d}{8}\pi^2 ab \quad A_2 = \frac{td^2}{16g}\pi^4 ab. \tag{5.18e, f}$$

$(U+V)$ is evidently a function of a_{mn}, λ and μ, there being a fresh trio of values for each mode m, n. If the plate is to be in equilibrium it is essential that $(U+V)$ should be stationary with respect to each of these unknown variables. For each mode, therefore, there are three equations to be satisfied:

$$\frac{\partial}{\partial \lambda}(U+V) = \frac{\partial}{\partial \mu}(U+V) = \frac{\partial}{\partial a_{mn}}(U+V) = 0. \tag{5.19}$$

These equations may be used to determine the values of a_{mn}, λ, μ in each separate mode. In fact, because the (m, n)th values of a_{mn}, λ, μ occur only in the (m, n)th mode, it is permissible to replace $(U+V)$ in equation (5.19) by $(U+V)_{mn}$ only.

It is convenient to begin by writing the total energy in the form:

$$(U+V)_{mn} = B_{xx}\lambda^2 + B_{yy}\mu^2 + 2B_{xy}\lambda\mu + 2B_x\lambda + 2B_y\mu + B_0, \tag{5.20}$$

where

$$B_{xx} = \left\{GA_1\frac{m^2}{a^2} + EA_2\left(\frac{m^4}{a^4} + \frac{1-\nu}{2}\frac{m^2n^2}{a^2b^2}\right)\right\}a_{mn}^2, \tag{5.21a}$$

$$B_{yy} = \left\{GA_1\frac{n^2}{b^2} + EA_2\left(\frac{n^4}{b^4} + \frac{1-\nu}{2}\frac{m^2n^2}{a^2b^2}\right)\right\}a_{mn}^2, \tag{5.21b}$$

$$B_{xy} = EA_2\left(\frac{1+\nu}{2}\right)\frac{m^2n^2}{a^2b^2}a_{mn}^2, \tag{5.21c}$$

$$B_x = -GA_1\frac{m^2}{a^2}a_{mn}^2 \quad B_y = -GA_1\frac{n^2}{b^2}a_{mn}^2, \tag{5.21d, e}$$

$$B_0 = GA_1\left(\frac{m^2}{a^2} + \frac{n^2}{b^2}\right)a_{mn}^2 + (V_1+V_2)_{mn}. \tag{5.21f}$$

Then, by (5.19),

$$\frac{1}{2}\frac{\partial}{\partial \lambda}(U+V) = \frac{1}{2}\frac{\partial(U+V)_{mn}}{\partial \lambda} = B_{xx}\lambda + B_{xy}\mu + B_x = 0, \quad (5.22a)$$

$$\frac{1}{2}\frac{\partial}{\partial \mu}(U+V) = \frac{1}{2}\frac{\partial(U+V)_{mn}}{\partial \mu} = B_{yy}\mu + B_{xy}\lambda + B_y = 0. \quad (5.22b)$$

If equations (5.22a) and (5.22b) are multiplied respectively by λ and μ and added, then

$$B_{xx}\lambda^2 + B_{yy}\mu^2 + 2B_{xy}\lambda\mu + B_x\lambda + B_y\mu = 0. \quad (5.23)$$

The *stationary* value of $(U+V)_{mn}$ is therefore obtained by equating this expression to zero in equation (5.20), leaving only:

$$(U+V)_{mn} = B_x\lambda + B_y\mu + B_0. \quad (5.24)$$

By solving (5.22a) and (5.22b) it is possible to show that in this particular problem the solution of equations (5.22) is such that $\mu = \lambda$.

If λ is substituted for μ in equation (5.22a), then the following result can be obtained for λ:

$$\lambda = -\frac{B_x}{B_{xx}+B_{xy}} = +\frac{1}{1+\varrho\Omega}, \quad (5.25a)$$

where
$$\Omega = \frac{m^2b^2}{a^2} + n^2, \quad (5.25b)$$

$$\varrho = \frac{\pi^2}{2g}\frac{E}{G}\frac{td}{b^2}. \quad (5.25c)$$

All three quantities λ, Ω and ϱ are non-dimensional; the first and second take different values for different modes; the third is constant and represents the product of π^2/b^2 and the ratio of the flexural rigidity $Etd^2/2g$ to the shear stiffness Gd.[†]

[†] The corresponding results which are obtained when the thickness of the faces is not neglected are given in equations (6.34b) and (6.35). The shear stiffness is then Gd^2/c rather than Gd.

Substitution of the value of λ from equation (5.25a) into equation (5.24) provides an expression for $(U+V)_{mn}$ as a function of a_{mn}:

$$(U+V)_{mn} = \frac{Gd}{8}\pi^2 a_{mn}^2 \frac{a}{b} \frac{\varrho\Omega^2}{1+\varrho\Omega} + \frac{N_x}{2}\pi^2 a_{mn}^2 \frac{ab}{4} \frac{m^2}{a^2}$$

$$-4q\frac{a_{mn}}{\pi^2}\frac{ab}{mn}. \qquad (5.26)$$

For equilibrium $(U+V)_{mn}$ must be stationary with respect to a_{mn}:

$$\frac{\partial}{\partial a_{mn}}(U+V)_{mn} = \left\{\frac{G\,d\pi^2}{4}\frac{a}{b}\frac{\varrho\Omega^2}{1+\varrho\Omega} + N_x\pi^2\frac{ab}{4}\frac{m^2}{a^2}\right\}a_{mn}$$

$$-\frac{4q}{\pi^2}\frac{ab}{mn} = 0. \qquad (5.27)$$

Suppose for the present that the transverse pressure q is zero. The problem now is to determine the critical value of N_x which causes the plate to buckle. If the plate buckles in the (m, n)th mode, a_{mn} is non-zero and equation (5.27) is satisfied only when the coefficient of a_{mn} is zero, that is, when

$$-N_x = \frac{Gd}{m^2}\left(\frac{a}{b}\right)^2 \frac{\varrho\Omega^2}{1+\varrho\Omega} = P_{xmn}. \qquad (5.28)$$

Equation (5.28) defines P_{xmn}, the critical edge load per unit length which causes buckling in the (m, n)th mode. It is necessary to determine the mode which gives the lowest critical load. Certainly for any given m, the smallest value of n gives the lowest critical load. Equation (5.28) may therefore be rearranged, with $n = 1$, as follows:

$$P_{xmn} = \frac{\pi^2 D_2}{b^2} K_1, \qquad (5.29a)$$

where D_2 = flexural rigidity of sandwich = $Etd^2/2g$. (5.29b)

$$K_1 = \frac{\{(mb/a)+(a/mb)\}^2}{1+\varrho[(m^2b^2/a^2)+1]}. \qquad (5.29c)$$

It has already been observed that ϱ (equation (5.25c)) is essentially the ratio of the flexural and shear rigidities. When the shear rigidity is infinite, ϱ vanishes and equation (5.29) is identical with the standard result for the buckling of a plate in the absence of shear deformations.[35,14]

Figure 5.4 shows the minimum value of K_1 plotted against a/b for various values of ϱ. For example, a family of curves is shown for $\varrho = 0$ and $m = 1, 2, 3 \ldots$. Only the lower envelope of the family is of practical interest. For this reason only the envelopes are shown of the families of curves for other values of ϱ.

In order to determine the buckling load it is necessary only to read K_1 from Fig. 5.4 and to insert the result in equation (5.29a).

If the plate is of infinite length in the x-direction it can be shown that, when $\varrho < 1$, K_1 has a true minimum value of $4/(1+\varrho)^2$ cor-

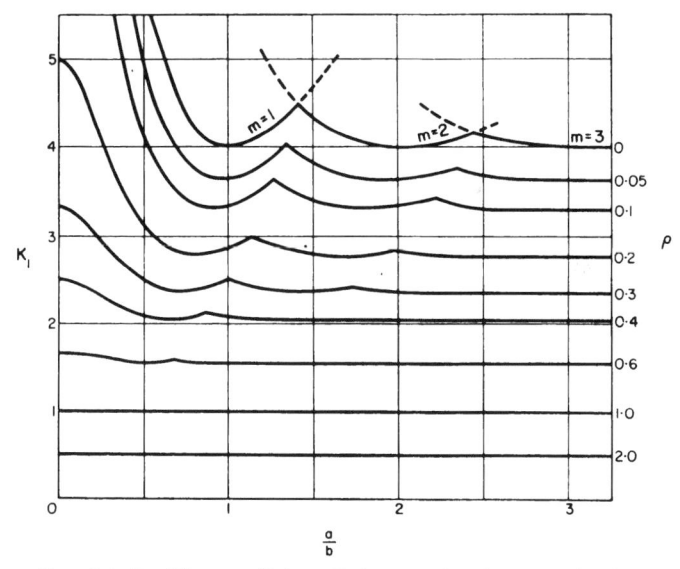

FIG. 5.4. Buckling coefficient K_1 in equation (5.29a). Simply-supported isotropic sandwich with thin faces and uniform edge load in the x-direction.

$$\left\{ \varrho = \frac{\pi^2}{2(1-\nu_f^2)} \frac{E_f}{G_c} \frac{td}{b^2} \right\}.$$

responding to $n = 1$ and a half-wavelength (a/m) in the x-direction equal to $b\sqrt{[(1-\varrho)/(1+\varrho)]}$. When $\varrho > 1$, the smallest value of K_1 is equal to $1/\varrho$, corresponding to $n = 1$ and an infinitely small half-wavelength in the x-direction.†

The equation for the (m, n)th critical load (5.28) may be written in an alternative form:

$$\frac{1}{P_{xmn}} = \frac{1}{P'} + \frac{1}{P''}. \tag{5.30}$$

Here P' is the (m, n)th critical load in the absence of core shear deformations and P'' may be described as the (m, n)th shear buckling load:

$$P' = \frac{\pi^2 D_2}{b^2}\left(\frac{mb}{a} + \frac{n^2 a}{mb}\right)^2, \tag{5.31a}$$

$$P'' = \frac{G\,da}{b}\left(\frac{b}{a} + \frac{n^2 a}{m^2 b}\right). \tag{5.31b}$$

Equation (5.31a) is, of course, a standard result.[35.14] Equations (5.29a) and (5.30) are analogous with equations (3.6a) and (3.6b) for a pin-ended strut with thin faces.

Suppose now that the edge load N_x is zero and that the plate supports a uniform transverse pressure q. After substitution for ϱ, equation (5.27) gives the amplitude of the (m, n)th mode:

$$a_{mn} = \frac{16qb^4}{\pi^6 mn\,D_2}\,\frac{1+\varrho\Omega}{\Omega^2} \quad (m, n \text{ odd}) \tag{5.32}$$

$$= 0 \text{ (otherwise)}.$$

When the core is rigid in shear, ϱ vanishes and equation (5.32) corresponds with the standard result for the bending of a plate in the absence of shear deformations.[35.3]

---

† This is virtually antisymmetrical wrinkling of the kind discussed in Chapter 8. In fact, the half-wavelength will be small but not zero because the faces have some local bending stiffness, which has been ignored in this chapter.

Equation (5.32) may also be written in the form

$$a_{mn} = \frac{16qb^4}{\pi^6 mn\, D_2 \Omega^2} + \frac{16qb^2}{\pi^4 mn Gd\Omega} \quad (m, n \text{ odd}) \quad (5.33)$$
$$= 0 \text{ (otherwise)}.$$

The first and second terms on the right-hand side represent the bending and shear deformations respectively. The ratio of the shear deformation to the bending deformation is

$$\varrho\Omega, \quad \text{or} \quad \frac{1-\lambda}{\lambda}, \quad \text{or} \quad \frac{\pi^2}{b^2} \frac{D_2}{Gd} \cdot \Omega.$$

Deflections are given by inserting the value of a_{mn} in equation (5.12). Peak deflections occur at the centre of the plate $x = a/2$, $y = b/2$, in which case

$$w_{max} = \frac{16qb^4}{\pi^6 D_2} \sum\sum \left\{ \frac{(-1)^{(m-1)/2}(-1)^{(n-1)/2}}{mn} \cdot \frac{1+\varrho\Omega}{\Omega^2} \right\}$$
$$(m, n \text{ odd}). \quad (5.34)$$

This may also be written, more conveniently,

$$w_{max} = \frac{qb^4}{D_2}(\beta_1 + \varrho\beta_2), \quad (5.35a)$$

where

$$\beta_1 = \frac{16}{\pi^6} \sum\sum \frac{(-1)^{(m-1)/2}(-1)^{(n-1)/2}}{mn\Omega^2} \quad (m, n \text{ odd}), \quad (5.35b)$$

$$\beta_2 = \frac{16}{\pi^6} \sum\sum \frac{(-1)^{(m-1)/2}(-1)^{(n-1)/2}}{mn\Omega} \quad (m, n \text{ odd}). \quad (5.35c)$$

Values of β_1 and β_2 may be obtained from Fig. 5.5 for plates with various a/b ratios.

The stresses in the faces and in the core may also be obtained from the result (5.32). For example, the direct stress in the faces,

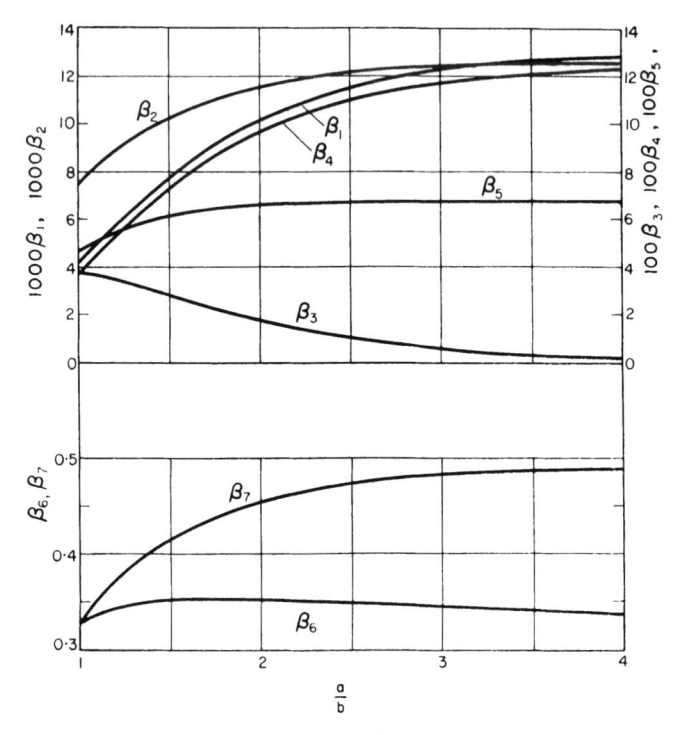

FIG. 5.5. Values of β_1–β_7 in equations (5.35) and (5.39). Isotropic panel with very thin faces. (The summations in equations (5.35) and (5.40) are taken up to and including the term $m = n = 23$.)

σ_x, is equal to $(E/g)(e_x + \nu e_y)$. The strains e_x and e_y are defined by equation (5.5) when $z = \mp d/2$. Hence:

$$\sigma_x = \pm \frac{E\,d\lambda}{2g}\left(\frac{\partial^2 w}{\partial x^2} + \nu\frac{\partial^2 w}{\partial y^2}\right). \tag{5.36a}$$

Similarly

$$\sigma_y = \pm \frac{E\,d\lambda}{2g}\left(\frac{\partial^2 w}{\partial y^2} + \nu\frac{\partial^2 w}{\partial x^2}\right). \tag{5.36b}$$

The shear stress τ_{xy} in the faces is equal to $\bigl(E/2(1+\nu)\bigr)\gamma_{xy}$ where the strain τ_{xy} is given by equation (5.6) when $z = \mp d/2$:

$$\tau_{xy} = \pm \frac{E\,d\lambda}{2(1+\nu)} \frac{\partial^2 w}{\partial x\,\partial y}. \tag{5.37}$$

The shear stress τ_{zx} in the zx-plane in the core is equal to $G\gamma_{zx}$, where the strain γ_{zx} is given by equation (5.1):

$$\tau_{zx} = G(1-\lambda)\frac{\partial w}{\partial x}. \tag{5.38a}$$

Similarly

$$\tau_{yz} = G(1-\lambda)\frac{\partial w}{\partial y}. \tag{5.38b}$$

Each of the stresses may be evaluated by appropriate differentiation of equation (5.12) and the insertion of the value of a_{mn} from equation (5.32). Usually the peak stresses will be of interest; it can be shown that the direct stresses in the faces are greatest at the centre of the plate ($x = a/2$, $y = b/2$); the shear stress in the faces is greatest at a corner (e.g. $x = 0$, $y = 0$); the core shear stress τ_{zx} is greatest in the middle of the sides of length b (e.g. $x = 0$, $y = b/2$); and the core shear stress τ_{yz} is greatest in the middle of the sides of length a (e.g. $x = a/2$, $y = 0$). The results may be summarized in the following form:

$$\sigma_x = \frac{qb^2}{dt}(\beta_3 + \nu\beta_4); \quad \sigma_y = \frac{qb^2}{dt}(\beta_4 + \nu\beta_3), \quad (5.39\text{a, b})$$

$$\tau_{xy} = \frac{qb^2}{dt}(1-\nu)\beta_5, \tag{5.39c}$$

$$\tau_{zx} = \frac{qb}{d}\beta_6, \qquad \tau_{yz} = \frac{qb}{d}\beta_7, \quad (5.39\text{ d, e})$$

where

$$\beta_3 = \sum\sum \frac{16}{\pi^4} \frac{(-1)^{(m-1)/2}(-1)^{(n-1)/2}}{\Omega^2} \frac{m}{n} \frac{b^2}{a^2}, \tag{5.40a}$$

$$\beta_4 = \sum\sum \frac{16}{\pi^4} \frac{(-1)^{(m-1)/2}(-1)^{(n-1)/2}}{\Omega^2} \frac{n}{m}, \tag{5.40b}$$

$$\beta_5 = \sum \sum \frac{16}{\pi^4} \frac{b}{a\Omega^2}, \qquad (5.40\text{c})$$

$$\beta_6 = \sum \sum \frac{16}{\pi^3} \frac{(-1)^{(n-1)/2}}{n\Omega} \frac{b}{a}, \qquad (5.40\text{d})$$

$$\beta_7 = \sum \sum \frac{16}{\pi^3} \frac{(-1)^{(m-1)/2}}{m\Omega}. \qquad (5.40\text{e})$$

In all cases, $m = 1, 3, 5, \ldots, n = 1, 3, 5\ldots$.

The functions β_3–β_7 are plotted against a/b in Fig. 5.5. It is of interest to observe that the expressions for the stresses do not include ϱ, nor any other term which refers to the shear stiffness of the core. Indeed, not only are the stresses independent of the shear stiffness of the core; it is easy to show that *the results in equations (5.39) and (5.40) are identical with those obtained when the core shear deformations are neglected.* The standard result for a simply supported rectangular plate with uniform transverse pressure, as given by Timoshenko [35.3] could therefore also be used to determine the stresses in a sandwich plate. Timoshenko's results are based on a more sophisticated analysis attributed to Levy; they give slightly more accurate results than Fig. 5.5 at high values of a/b, but the difference is not more than a few percent at the most. The lower accuracy of the Navier solution used here is due to the slow convergence of the series for the β-functions, especially β_6 and β_7;[†] the convergence deteriorates as a/b diverges from unity. For example, Timoshenko gives $\beta_6 = \beta_7 = 0.338$ at $a/b = 1$, with β_6 rising to 0.37 for $a/b > 2$ and with $\beta_7 = 0.5$ for $a/b > 4$. Figure 5.5 is derived from the summation of all terms up to and including $m = 23, n = 23$.

When the uniform transverse pressure q and the compressive edge load $P(= -N_x)$ per unit length act simultaneously, the value of a_{mn} may again be obtained from equation (5.27). After

---

[†] See Section 4.6 for a detailed illustration of the relatively slow convergence of the series for core shear stress in a beam.

some rearrangement, using the expression (5.28) for the (m, n)th critical load, the amplitude may be written in the following form:

$$a_{mn} = \frac{(a_{mn})_0}{1 - P/P_{xmn}}, \qquad (5.41)$$

where $(a_{mn})_0$, the amplitude when P is zero, is given by equation (5.32), and P_{xmn} by equation (5.29). The practical effect of the end load is to factor each term in the series for the β-functions by an amplification factor, $(1 - P/P_{xmn})^{-1}$. Because P_{xmn} incorporates the ratio ϱ, the stresses in the plate are no longer independent of the shear stiffness.

The method of this section, in which the energy of the system is minimized with respect to the amplitude of an assumed mode of deformation and to a parameter such as λ appears to have been first used for sandwich plates in refs. 3.1 and 3.2; the same procedure was used in ref. 1.5 and a number of later reports of the U.S. Forest Products Laboratory.

5.5 Other Types of Simply-supported Sandwich Plate

The analysis of Section 5.4 is not adequate for sandwich panels in which the faces and core are not isotropic. For example, it is often inappropriate when the faces are made of plywood or reinforced plastics, or when the core is made of certain honeycombs or expanded plastics which are stiffer in one direction than in another. It is quite practicable to extend the analysis to deal with such orthotropic panels, as in Chapter 6, but it is sometimes expedient to rely instead on the alternative method described in Chapter 7. This is because the notation of Chapter 7 lends itself not only to the analysis of panels with weak cores (in the sense that the cores contribute little to the flexural rigidity), but also to the analysis of panels with corrugated cores.

A corrugated-core sandwich panel is illustrated diagrammatically in Fig. 5.6. The corrugated core is often stiff enough to

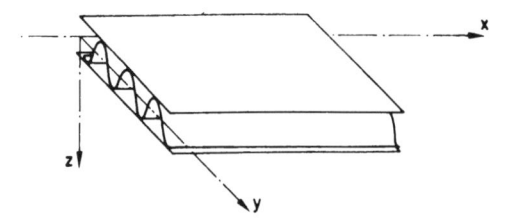

FIG. 5.6. Corrugated-core sandwich.

make a distinct contribution to the flexural rigidity in the zx-plane but not in the yz-plane. Consequently the usual assumption that the flexural rigidity of the core is negligible breaks down for bending in the zx-plane and the shear stresses in that plane can no longer be assumed constant over the depth of the core. Fortunately for the simplicity of the analysis, the shear stiffness of the corrugated core can usually be taken as infinite in the zx-plane, equivalent to writing $\lambda = 1$ in equation (5.1). However, if the analysis of Section 5.3 is to be applied to the corrugated-core panel, it is necessary to introduce additional terms to represent the strain energy associated with the zx-bending and possibly the twisting of the corrugated core. To do this would be to destroy the simplicity of the presentation.

On the other hand, the method of Chapter 7, although mathematically equivalent, uses a notation which permits the corrugated-core sandwich to be treated as merely a special case of a general orthotropic sandwich.

The method of Chapter 5, unlike that of Chapter 7, can be extended without much difficulty to deal with sandwich panels with thick faces. The extended technique was used in the analysis of beam columns with thick faces in Chapter 4. The extension to panels with thick faces is carried out in Chapter 6, in which the opportunity is taken to deal also with faces of unequal thickness and dissimilar materials. Section 6.7 contains the special case of an isotropic panel with identical thick faces, for comparison with the results of Chapter 5.

BENDING AND BUCKLING OF ISOTROPIC SANDWICH PANELS 97

5.6. Boundary Conditions for a Simply-supported Panel†

In a panel which is simply supported on four sides it might be reasonable to expect the following boundary conditions to be satisfied:

(i) $x = 0$ or a and $y = 0$ or b: $w = 0$
(ii) $x = 0$ or a : $\sigma_x = 0$
(iii) $y = 0$ or b : $\sigma_y = 0$ At $z = \pm \dfrac{d}{2}$.
(iv) $x = 0$ or a : $\tau_{xy} = 0$
(v) $y = 0$ or b : $\tau_{xy} = 0$

Condition (i) is completely satisfied by the sinusoidal displacement function (5.12).

Conditions (ii) to (v) express the fact that around their edges the faces should be free of in-plane stresses associated with the bending of the panel. They are equivalent to the requirements:

(iia) $x = 0$ or a: $M_x = 0$,
(iiia) $y = 0$ or b: $M_y = 0$,
(iva) $x = 0$ or a: $M_{xy} = 0$,
(va) $y = 0$ or b: $M_{yx} = 0$.

Condition (ii) may be written in terms of the displacement w by the the use of equation (5.5):

$$\sigma_x = \frac{E}{g}(e_x + \nu e_y) = \pm \frac{E}{g}\left(\frac{\partial^2 w}{\partial x^2} + \nu \frac{\partial^2 w}{\partial y^2}\right)\frac{\lambda d}{2}.$$

It can be seen from this that the sinusoidal displacement function (5.12) satisfies condition (ii). Similarly it can be shown to satisfy (iii).

Condition (iv) may be written in terms of the displacement w by the use of equation (5.6):

$$\tau_{xy} = G_f \gamma_{xy} = \pm G_f\, d\lambda\, \frac{\partial^2 w}{\partial x\, \partial y}.$$

† See Section 9.3 for further comments on boundary conditions.

It can be seen from this that the sinusoidal displacement function (5.12) does *not* satisfy conditions (iv) and (v).

Fortunately it *can* be shown to satisfy the following conditions instead (see equations (5.2) and (5.4)):

(vi) $x = 0$ or a: $\quad v = \pm d\mu \dfrac{\partial w}{\partial y} = 0,$

(vii) $y = 0$ or b: $\quad u = \pm d\lambda \dfrac{\partial w}{\partial x} = 0.$

Condition (vi) is equivalent to the statement that the core shear strain γ_{yz} is zero along the edges $x = 0$ or a. This corresponds to the presence of stiffeners along these edges. The stiffeners must connect the faces together to prevent relative sliding in the y-direction; on the other hand they should be flexible in torsion to avoid interference with the bending of the plate.

A change from boundary conditions (iv) and (v) to boundary conditions (vi) and (vii) has but slight effect on the state of stress in the panel except in the unlikely event of a system of loads which tends to rotate the faces bodily in opposite directions in the xy-plane. In any case, the conditions (vi) and (vii) are often the more realistic.

CHAPTER 6

BENDING AND BUCKLING OF ORTHOTROPIC SANDWICH PANELS WITH THICK DISSIMILAR FACES (RITZ METHOD)

6.1. Introduction

The bending and buckling of sandwich plates with identical very thin faces was considered in Chapter 5. In this chapter a more general analysis will be presented for sandwich panels in which the faces on opposite sides of the panel are of different thicknesses and different materials. This more general analysis is not confined to faces of negligible thickness but makes allowance for the effect of face thickness on the geometry of the deformation and for the local bending stiffness of the faces.

The analyses of Chapters 5 and 6 share the assumption that there is no stretching of the plate during bending and that the core makes no contribution to the flexural rigidity of the sandwich. As usual, this is taken to imply that shear stresses are constant through the depth of the core.

The number of variables involved in the following analysis is so large that it is not convenient here to provide formulae or graphs for the critical loads, stresses and transverse displacements in the general case. Instead a procedure will be given for the numerical evaluation of the desired information for a plate of stated dimensions and materials.

The special case of a sandwich with identical isotropic thick faces and an "isotropic" core is considered in detail in Section 6.7 in order to assess the differences between the "thick face analysis" of this chapter and the "thin face analysis" of Chapter 5.

Attention will be given mainly to simply-supported panels with uniform transverse pressure and with edge loads parallel with the x-axis, but the form of the analysis will be such that it can be adapted to deal with other types of load and edge conditions provided suitable modes of deformation can be selected.

The method follows the general principles established by Ericksen and March.[1.7, 1.8] Its relationship to other methods described in the literature on sandwich panels is discussed in Sections 9.1 and 9.2. The influences of factors such as different boundary conditions, special types of edge loading, large deflections and initial deformations are mentioned briefly in the later parts of Chapter 9. Sources of additional information for design purposes are listed in Chapter 10.

6.2. Displacements and Strains

The interface between the core and the upper face lies in the xy-plane (Fig. 6.1). The thicknesses of the upper face, core and lower face are t_1, c, t_2 respectively. In Fig. 6.2 is shown a section in the zx-plane. The sandwich has deflected under load and the

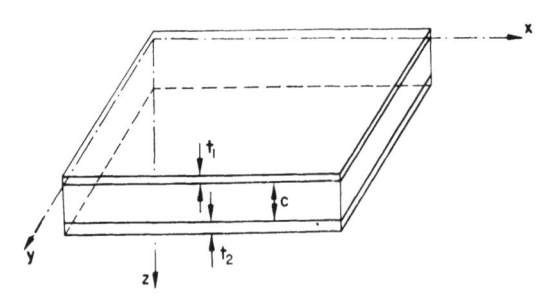

FIG. 6.1. Sandwich panel with thick unequal faces.

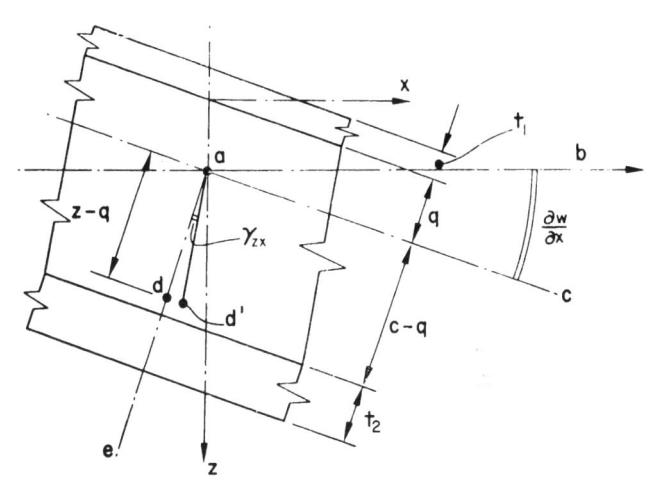

FIG. 6.2. Section through deflected sandwich panel in the zx-plane.

point (a) has descended a distance w in the yz-plane. The neutral axis (on which there are no x-displacements) has descended to ab and the rotated through an angle $\partial w/\partial x$ to ac. The normal to the neutral axis has rotated to ae. As a result of shear deformation in the core, a point such as d (on the normal ae) moves to d', where the angle dad' is the shear strain γ_{zx}. The point d is a distance z below the upper interface and, if the distance from the interface to the neutral axis is denoted by q, then $ad' = z - q$. Also, if the rotation $d'az$ of the line ad' is denoted by $\lambda(\partial w/\partial z)$,† then the displacement of d' in the x-direction is given by

$$u = -\lambda(z-q)\frac{\partial w}{\partial x}. \tag{6.1a}$$

The parameter λ may take any value between the limits $+1$ (for a core rigid in shear) and $-(t_1+t_2)/2c$ (for a core completely flexible in shear). The value of q is to be determined.

† In effect this is an alternative definition of λ.

A similar diagram may be drawn for the yz-plane, from which it may be deduced that the displacement in the y-direction is

$$v = -\mu(z-r)\frac{\partial w}{\partial y}. \tag{6.1b}$$

Here μ is a parameter such that a line in the core (originally vertical) rotates in the yz-plane through an angle $\mu(\partial w/\partial y)$. The quantity r is the distance from the upper core/face interface to the neutral axis in the y-direction (on which there are no y-displacements).

The parameters λ and μ are analogous with the ones used in the previous chapter. The values q and r are identical when the faces are made of similar materials.

The first step is to write down the displacements in the core and in the faces and then, by differentiation, to evaluate the strains. It is important to note that although, in general, λ and μ are functions of x and y, it is assumed here for the purpose of differentiation that λ and μ are constants. This will later be shown to be valid for certain common problems (Chapter 7).

Core displacements

These are given by equations (6.1a, b).

The relationships which follow are derived from equations (6.1) and from the geometry of Fig. 6.2.

Face displacements

Upper face:

$$u = -\lambda(0-q)\frac{\partial w}{\partial x} - z\frac{\partial w}{\partial x} = (\lambda q - z)\frac{\partial w}{\partial x}, \tag{6.2a}$$

$$v = -\mu(0-r)\frac{\partial w}{\partial y} - z\frac{\partial w}{\partial y} = (\mu r - z)\frac{\partial w}{\partial y}. \tag{6.2b}$$

Lower face:

$$u = -\lambda(c-q)\frac{\partial w}{\partial x} - (z-c)\frac{\partial w}{\partial x} = -[\lambda(c-q)+z-c]\frac{\partial w}{\partial x}, \tag{6.3a}$$

$$v = -\mu(c-r)\frac{\partial w}{\partial y} - (z-c)\frac{\partial w}{\partial y} = -[\mu(c-r)+z-c]\frac{\partial w}{\partial y}.$$
(6.3b)

Core strains

$$\gamma_{zx} = \frac{\partial u}{\partial z} + \frac{\partial w}{\partial x} = -\lambda\frac{\partial w}{\partial x} + \frac{\partial w}{\partial x} = (1-\lambda)\frac{\partial w}{\partial x}, \quad (6.4a)$$

$$\gamma_{yz} = \frac{\partial v}{\partial z} + \frac{\partial w}{\partial y} = -\mu\frac{\partial w}{\partial y} + \frac{\partial w}{\partial y} = (1-\mu)\frac{\partial w}{\partial y}. \quad (6.4b)$$

γ_{xy} is not required.

Upper face membrane strains $(z = -t_1/2)$

$$e_x = \frac{\partial u}{\partial x} = \left(\lambda q + \frac{t_1}{2}\right)\frac{\partial^2 w}{\partial x^2}, \quad (6.5a)$$

$$e_y = \frac{\partial v}{\partial y} = \left(\mu r + \frac{t_1}{2}\right)\frac{\partial^2 w}{\partial y^2}, \quad (6.5b)$$

$$\gamma_{xy} = \frac{\partial u}{\partial y} + \frac{\partial v}{\partial x} = (\lambda q + \mu r + t_1)\frac{\partial^2 w}{\partial x\,\partial y}, \quad (6.5c)$$

$$\gamma_{zx} = \gamma_{yz} = \sigma_z = 0.$$

Lower face membrane strains $(z = c + t_2/2)$

$$e_x = \frac{\partial u}{\partial x} = -\left[\lambda(c-q) + \frac{t_2}{2}\right]\frac{\partial^2 w}{\partial x^2}, \quad (6.6a)$$

$$e_y = \frac{\partial v}{\partial y} = -\left[\mu(c-r) + \frac{t_2}{2}\right]\frac{\partial^2 w}{\partial y^2}, \quad (6.6b)$$

$$\gamma_{xy} = \frac{\partial u}{\partial y} + \frac{\partial v}{\partial x} = -[\lambda(c-q) + \mu(c-r) + t_2]\frac{\partial^2 w}{\partial x\,\partial y}, \quad (6.6c)$$

$$\gamma_{zx} = \gamma_{yz} = \sigma_z = 0.$$

Upper face bending strains

$$e_x = -\left(z+\frac{t_1}{2}\right)\frac{\partial^2 w}{\partial x^2}; \quad e_y = -\left(z+\frac{t_1}{2}\right)\frac{\partial^2 w}{\partial y^2};$$

$$\gamma_{xy} = -2\left(z+\frac{t_1}{2}\right)\frac{\partial^2 w}{\partial x\,\partial y}. \quad (6.7\text{ a, b, c})$$

Lower face bending strains

$$e_x = -\left(z-c-\frac{t_2}{2}\right)\frac{\partial^2 w}{\partial x^2}, \quad (6.8a)$$

$$e_y = -\left(z-c-\frac{t_2}{2}\right)\frac{\partial^2 w}{\partial y^2}, \quad (6.8b)$$

$$\gamma_{xy} = -2\left(z-c-\frac{t_2}{2}\right)\frac{\partial^2 w}{\partial x\,\partial y}. \quad (6.8c)$$

6.3. Strain Energy

The strain energy U per unit volume of an orthotropic elastic solid in which σ_z is zero is given by:

$$U = \{E_x e_x^2 + E_y e_y^2 + 2E_x \nu_y e_x e_y\}/2g + \{G_{xy}\gamma_{xy}^2 + G_{yz}\gamma_{yz}^2 + G_{zx}\gamma_{zx}^2\}/2, \quad (6.9)$$

where E_x, E_y, G_{yz}, G_{zx} are the various elastic and shear moduli, ν_x and ν_y are the Poisson's ratios and g is equal to $(1-\nu_x\nu_y)$. When the only stress present is σ_x, then ν_x is defined as the ratio of the strains in the y-and x-directions, with a negative sign. Further details of equation (6.9) and the properties of orthotropic materials are given in Appendix I.

Shear strain energy in the core

For the core all the elastic moduli and G_{xy} are zero. The shear strains from equation (6.4) may be substituted in equation (6.9) as follows:

$$U_c = \frac{1}{2}\int_V \left\{ G_{zx}(1-\lambda)^2 \left(\frac{\partial w}{\partial x}\right)^2 + G_{yz}(1-\mu)^2 \left(\frac{\partial w}{\partial y}\right)^2 \right\} dV$$

$$= \frac{c}{2}\int_A \left\{ G_{zx}(1-\lambda)^2 \left(\frac{\partial w}{\partial x}\right)^2 + G_{yz}(1-\mu)^2 \left(\frac{\partial w}{\partial y}\right)^2 \right\} dA. \quad (6.10)$$

Here $\int_V dV$ and $\int_A dA$ indicate integration over the volume of the core and the area of the plate respectively.

Membrane strain energy in upper face

These stresses and strains are constant through the thickness t_1 of the face. Strains e_x, e_y, γ_{xy} are given by equations (6.5); strains γ_{zx}, γ_{zy} are zero.

$$U_{fm1} = \frac{1}{2g}\int_V \left\{ E_x\left(\lambda q + \frac{t_1}{2}\right)^2 \left(\frac{\partial^2 w}{\partial x^2}\right)^2 + E_y\left(\mu r + \frac{t_1}{2}\right)^2 \left(\frac{\partial^2 w}{\partial y^2}\right)^2 \right.$$

$$\left. + 2E_x\nu_y\left(\lambda q + \frac{t_1}{2}\right)\left(\mu r + \frac{t_1}{2}\right)\frac{\partial^2 w}{\partial x^2}\frac{\partial^2 w}{\partial y^2} \right\} dV$$

$$+ \frac{1}{2}\int_V G_{xy}(\lambda q + \mu r + t_1)^2 \left(\frac{\partial^2 w}{\partial x \partial y}\right)^2 dV. \quad (6.11a)$$

This may be integrated across the thickness of the face.

$$U_{fm1} = \frac{t_1}{2g}\int_A \left\{ E_x\left(\lambda q + \frac{t_1}{2}\right)^2 \left(\frac{\partial^2 w}{\partial x^2}\right)^2 + E_y\left(\mu r + \frac{t_1}{2}\right)^2 \left(\frac{\partial^2 w}{\partial y^2}\right)^2 \right.$$

$$\left. + 2E_x\nu_y\left(\lambda q + \frac{t_1}{2}\right)\left(\mu r + \frac{t_1}{2}\right)\frac{\partial^2 w}{\partial x^2} \cdot \frac{\partial^2 w}{\partial y^2} \right\} dA$$

$$+ \frac{t_1}{2}\int_A G_{xy}(\lambda q + \mu r + t_1)^2 \left(\frac{\partial^2 w}{\partial x \partial y}\right)^2 dA. \quad (6.11b)$$

Membrane strain energy in lower face

Similarly,

$$\begin{aligned}U_{fm2} = \frac{t_2}{2g} \int_A &\left\{ E_x \left[\lambda(c-q) + \frac{t_2}{2} \right]^2 \left(\frac{\partial^2 w}{\partial x^2} \right)^2 \right.\\ &+ E_y \left[\mu(c-r) + \frac{t_2}{2} \right]^2 \left(\frac{\partial^2 w}{\partial y^2} \right)^2 \\ &\left. + 2E_x \nu_y \left[\lambda(c-q) + \frac{t_2}{2} \right]\left[\mu(c-r) + \frac{t_2}{2} \right] \frac{\partial^2 w}{\partial x^2} \cdot \frac{\partial^2 w}{\partial y^2} \right\} dA \\ &+ \frac{t_2}{2} \int_A G_{xy} [\lambda(c-q) + \mu(c-r) + t_2]^2 \left(\frac{\partial^2 w}{\partial x \partial y} \right)^2 dA. \quad (6.12)\end{aligned}$$

Bending strain energy in upper face

The non-zero strains are given by equations (6.7). Substitution in equation (6.9) and integration over the thickness of the skin gives the strain energy of bending:

$$\begin{aligned}U_{fb1} = \frac{t_1^3}{24g} \int_A &\left\{ E_x \left(\frac{\partial^2 w}{\partial x^2} \right)^2 + E_y \left(\frac{\partial^2 w}{\partial y^2} \right)^2 + 2E_x \nu_y \frac{\partial^2 w}{\partial x^2} \frac{\partial^2 w}{\partial y^2} \right\} dA \\ &+ \frac{t_1^3}{6} \int_A G_{xy} \left(\frac{\partial^2 w}{\partial x \partial y} \right)^2 dA. \quad (6.13)\end{aligned}$$

Bending strain energy in lower face

Similarly,

$$\begin{aligned}U_{fb2} = \frac{t_2^3}{24g} \int_A &\left\{ E_x \left(\frac{\partial^2 w}{\partial x^2} \right)^2 + E_y \left(\frac{\partial^2 w}{\partial y^2} \right)^2 + 2E_x \nu_y \frac{\partial^2 w}{\partial x^2} \frac{\partial^2 w}{\partial y^2} \right\} dA \\ &+ \frac{t_2^3}{6} \int_A G_{xy} \left(\frac{\partial^2 w}{\partial x \partial y} \right)^2 dA. \quad (6.14)\end{aligned}$$

6.4. Potential Energy of Applied Forces

Suppose that the plate is subjected to a tensile edge load N_x per unit length in the x-direction along the edges $x = 0, a$, to a tensile edge load N_y per unit length in the y-direction along the edges $y = 0, b$, and to a uniformly distributed transverse load q per unit area. As the plate bends, the edges $x = 0, a$, approach each other and the forces N_x gain potential energy. Similarly, the edges $y = 0, b$, approach each other and the forces N_y gain potential energy. Also, the forces q lose potential energy as they descend in the z-direction. The changes of potential energy associated with N_x and N_y and with q are

$$V_1 = \frac{N_x}{2} \int_A \left(\frac{\partial w}{\partial x}\right)^2 dA + \frac{N_y}{2} \int_A \left(\frac{\partial w}{\partial y}\right)^2 dA, \qquad (6.15)$$

$$V_2 = -q \int_A w \, dA. \qquad (6.16)$$

A fuller discussion of the derivation of these expressions is given in Sections 4.3 and 5.3.

6.5. Minimization of Total Energy

It will be assumed that the deflection of the plate may be expressed in the form

$$w = \sum_{m=1}^{\infty} \sum_{n=1}^{\infty} a_{mn} \phi_m(x) \psi_n(y). \qquad (6.17)$$

In this equation $\phi_m(x)$, $\psi_n(y)$ are assumed functions of x and y respectively. For example, sinusoidal functions would be appropriate for a simply-supported plate, the boundary conditions for which are discussed in Section 5.6.

The quantities λ, μ, q, r have different values in the different modes m, n. For example, the core shear strain γ_{zx} in the m, nth mode is given by

$$(\gamma_{zx})_{mn} = (1-\lambda_{mn})\frac{\partial w_{mn}}{\partial x},$$

where $w_{mn} = a_{mn}\phi_m(x)\,\psi_n(y)$. The strain energy associated with the total shear strain γ_{zx} is therefore

$$\frac{cG_{zx}}{2}\int_A\left\{\sum_{m=1}^{\infty}\sum_{n=1}^{\infty}(1-\lambda_{mn})\frac{\partial w_{mn}}{\partial x}\right\}^2 dA.$$

This corresponds to the first term in equation (6.10) and it may be expanded as follows:

$$\frac{cG_{zx}}{2}\int_A\left\{\sum\sum(1-\lambda_{mn})\,a_{mn}\frac{\partial\phi_m}{\partial x}\,\psi_n\right\}^2 dA$$

$$= \frac{cG_{zx}}{2}\int_A\left\{\sum\sum(1-\lambda_{mn})^2\,a_{mn}^2\left(\frac{\partial\phi_m}{\partial x}\right)^2\psi_n^2\right.$$

$$\left. + 2\sum\sum(1-\lambda_{mn})\,a_{mn}\frac{\partial\phi_m}{\partial x}\,\psi_n(1-\lambda_{pq})\,a_{pq}\frac{\partial\phi_p}{\partial x}\,\psi_q\right\}dA.$$

$$(m \neq p, \quad n \neq q)$$

If the functions ϕ, ψ have suitable orthogonal properties the second term in the last expression vanishes, leaving

$$\frac{cG_{zx}}{2}\sum_{m=1}^{\infty}\sum_{n=1}^{\infty}\left\{(1-\lambda_{mn})^2\,a_{mn}^2\int_A\left(\frac{\partial\phi_m}{\partial x}\right)^2\psi_n^2\,dA\right\}$$

Similar operations may be performed on all the integrals found in the expressions for the various strain energies and potential energies in the preceding sections. The work which follows will be simplified very considerably by the introduction of the following notation:

$$\int_A \left(\frac{\partial^2 \phi_m}{\partial x^2}\right)^2 \psi_n^2 \, dA = (i_1)_{mn} \qquad \int_A \phi_m^2 \left(\frac{\partial^2 \psi_n}{\partial y^2}\right)^2 dA = (i_2)_{mn},$$

$$\int_A \left(\frac{\partial^2 \phi_m}{\partial x^2} \psi_n\right) \left(\phi_m \frac{\partial^2 \psi_n}{\partial y^2}\right) dA = (i_3)_{mn}$$

$$\int_A \left(\frac{\partial \phi_m}{\partial x} \cdot \frac{\partial \psi_n}{\partial y}\right)^2 dA = (i_4)_{mn}$$

$$\int_A \left(\frac{\partial \phi_m}{\partial x}\right)^2 \psi_n^2 \, dA = (i_5)_{mn} \qquad \int_A \phi_m^2 \left(\frac{\partial \psi_n}{\partial y}\right)^2 dA = (i_6)_{mn},$$

$$\int_A \phi_m \psi_n \, dA = (i_7)_{mn}. \qquad (6.18)$$

Values of the integrals $i_1 - i_7$ for functions ϕ, ψ appropriate to plates with simply-supported boundary conditions only are given in Table 6.1. The total energy of the system is the sum of the energies given by equations (6.10)–(6.16):

$$U + V = U_c + U_{fm1} + U_{fm2} + U_{fb1} + U_{fb2} + V_1 + V_2$$

$$= \sum_{m=1}^{\infty} \sum_{n=1}^{\infty} \left\{ a_{mn}^2 f(\lambda_{mn}, \mu_{mn}, \lambda_{mn} q_{mn}, \mu_{mn} r_{mn}) \right.$$

$$\left. + \frac{N_x}{2} a_{mn}^2 (i_5)_{mn} + \frac{N_y}{2} a_{mn}^2 (i_6)_{mn} - q a_{mn} (i_7)_{mn} \right\}. \quad (6.19)$$

It is convenient to introduce the notation:

$$(b_1)_{mn} = \frac{E_x}{g}(i_1)_{mn}; \quad (b_2)_{mn} = \frac{E_y}{g}(i_2)_{mn}; \quad (b_3)_{mn} = \frac{E_x \nu_y}{g}(i_3)_{mn};$$
$$(6.20)$$
$$(b_4)_{mn} = G_{xy}(i_4)_{mn}; \quad (b_5)_{mn} = G_{zx}(i_5)_{mn}; \quad (b_6)_{mn} = G_{yz}(i_6)_{mn}.$$

The upper and lower faces may be made of different materials, in which case there will be two sets of quantities b_1–b_4, one for each face. G_{zx} and G_{yz} refer to the core material, the others to the face materials.

The form of the function f may now be determined from equations (6.10)–(6.14) as follows. The suffixes mn will be dropped from λ, μ, q, r and b_1–b_6 to avoid confusion:

$$f(\lambda_{mn}, \mu_{mn}, \lambda_{mn}q_{mn}, \mu_{mn}r_{mn}) = \frac{c}{2}\{(1-\lambda)^2 b_5 + (1-\mu)^2 b_6\}$$

$$+ \frac{t_1}{2}\left\{\left(\lambda q + \frac{t_1}{2}\right)^2 b_1 + \left(\mu r + \frac{t_1}{2}\right)^2 b_2 + 2\left(\lambda q + \frac{t_1}{2}\right)\left(\mu r + \frac{t_1}{2}\right) b_3\right.$$

$$+ (\lambda q + \mu r + t_1)^2 b_4 \Big\} + \frac{t_2}{2}\left\{\left(\lambda c - \lambda q + \frac{t_2}{2}\right)^2 b_1 + \left(\mu c - \mu r + \frac{t_2}{2}\right)^2 b_2\right.$$

$$+ 2\left(\lambda c - \lambda q + \frac{t_2}{2}\right)\left(\mu c - \mu r + \frac{t_2}{2}\right) b_3 + (\lambda c - \lambda q + \mu c - \mu r + t_2)^2 b_4\Big\}$$

$$+ \frac{t_1^3}{24}\{b_1 + b_2 + 2b_3 + 4b_4\} + \frac{t_2^3}{24}\{b_1 + b_2 + 2b_3 + 4b_4\}. \quad (6.21)$$

TABLE 6.1. VALUES OF INTEGRALS IN EQUATION (6.18) FOR A SIMPLY-SUPPORTED PANEL

$$\phi_m = \sin\frac{m\pi x}{a}; \quad \psi_n = \sin\frac{n\pi y}{b}$$

$i_1 = \dfrac{\pi^4 m^4}{a^4} \cdot \dfrac{ab}{4}$	$i_2 = \dfrac{\pi^4 n^4}{b^4} \cdot \dfrac{ab}{4}$

$$i_3 = \frac{\pi^4 n^2 m^2}{a^2 b^2} \cdot \frac{ab}{4} = i_4$$

$i_5 = \dfrac{\pi^2 m^2}{a^2} \cdot \dfrac{ab}{4}$	$i_6 = \dfrac{\pi^2 n^2}{b^2} \cdot \dfrac{ab}{4}$

$i_7 = \dfrac{4ab}{mn\pi^2}$ (m, n both odd) otherwise zero

The terms involving t_1 and t_2 refer to the upper and lower faces respectively and the appropriate values of b_1–b_4 must be used. The first term is associated with shear strain in the core; the second and third terms with membrane strains in the faces; the fourth and fifth terms with local bending strains in the faces.

For equilibrium it is necessary that[†] $(U+V)$ should be stationary with respect to λ_{mn}, μ_{mn}, $\lambda_{mn}q_{mn}$, $\mu_{mn}r_{mn}$ for all m, n. Inspection of equation (6.19) shows that this is true provided $f(\lambda_{mn}\ldots)$ is also stationary with respect to the same variables. The consequences of this conclusion may be explored more easily by writing

$$f(\lambda_{mn}\ldots) = f(x_1, x_2, x_3, x_4)$$
$$= \sum c_{ii}x_i^2 + 2\sum_{j>i} c_{ij}x_ix_j + 2\sum c_i x_i + c_0, \quad (6.22)$$

where $i, j = 1, 2, 3, 4$ and $x_1 = \lambda_{mn}$,
$$x_2 = \mu_{mn},$$
$$x_3 = \lambda_{mn}q_{mn},$$
$$x_4 = \mu_{mn}r_{mn}.$$

For $f(\ )$ to be stationary the following conditions must be satisfied:

$$\frac{1}{2}\frac{\partial f}{\partial x_1} = c_{11}x_1 + c_{12}x_2 + c_{13}x_3 + c_{14}x_4 + c_1 = 0,$$

$$\frac{1}{2}\frac{\partial f}{\partial x_2} = c_{12}x_1 + c_{22}x_2 + c_{23}x_3 + c_{24}x_4 + c_2 = 0,$$

$$\frac{1}{2}\frac{\partial f}{\partial x_3} = c_{13}x_1 + c_{23}x_2 + c_{33}x_3 + c_{34}x_4 + c_3 = 0,$$

$$\frac{1}{2}\frac{\partial f}{\partial x_4} = c_{14}x_1 + c_{24}x_2 + c_{34}x_3 + c_{44}x_4 + c_4 = 0.$$

(6.23)

† But not sufficient. See also equation (6.30a).

These equations may be multiplied respectively by x_1, x_2, x_3 and x_4 and then added to give the following result:

$$\sum c_{ii} x_i^2 + 2 \sum_{j>i} c_{ij} x_i x_j + \sum c_i x_i = 0. \tag{6.24}$$

When $f(\ )$ is stationary, this part of equation (6.22) vanishes, leaving only

$$f_{\min} = c_0 + \sum c_i x_i. \tag{6.25}$$

The solution of the equations (6.23) may be written in the form

$$x_1 = -\frac{\Delta_1}{\Delta}, \quad x_2 = -\frac{\Delta_2}{\Delta}, \quad x_3 = -\frac{\Delta_3}{\Delta}, \quad x_4 = -\frac{\Delta_4}{\Delta},$$

$$\tag{6.26a}$$

where $\Delta = \begin{vmatrix} c_{11} & c_{12} & c_{13} & c_{14} \\ c_{12} & c_{22} & c_{23} & c_{24} \\ c_{13} & c_{23} & c_{33} & c_{34} \\ c_{14} & c_{24} & c_{34} & c_{44} \end{vmatrix}, \quad \Delta_1 = \begin{vmatrix} c_1 & c_{12} & c_{13} & c_{14} \\ c_2 & c_{22} & c_{23} & c_{24} \\ c_3 & c_{23} & c_{33} & c_{34} \\ c_4 & c_{24} & c_{34} & c_{44} \end{vmatrix}$

etc. $\tag{6.26b}$

These results may be inserted in equation (6.25) to give finally the minimum value of $f(\ )$:

$$f_{\min} = c_0 - \frac{1}{\Delta} \sum c_i \Delta_i \quad i=1, 2, 3, 4. \tag{6.27}$$

The procedure just described is much simpler than the more obvious process of substituting the solution of equations (6.23) directly into equation (6.22).

The explicit values of the quantities c_0, c_i, c_{ij} are listed below for reference.

Suffixes m, n are omitted throughout.

$$c_{11} = \frac{c}{2} b_5 + \frac{c^2 t_2}{2} (b_1 + b_4)$$

$$c_{22} = \frac{c}{2} b_6 + \frac{c^2 t_2}{2} (b_2 + b_4)$$

$$c_{33} = \frac{t_1}{2} (b_1 + b_4) + \frac{t_2}{2} (b_1 + b_4)$$

$$c_{44} = \frac{t_1}{2} (b_2 + b_4) + \frac{t_2}{2} (b_2 + b_4)$$

$$2c_{12} = c^2 t_2 (b_3 + b_4)$$

$$2c_{13} = -c t_2 (b_1 + b_4)$$

$$2c_{14} = -c t_2 (b_3 + b_4)$$

$$2c_{23} = -c t_2 (b_3 + b_4)$$

$$2c_{24} = -c t_2 (b_2 + b_4)$$

$$2c_{34} = t_1 (b_3 + b_4) + t_2 (b_3 + b_4)$$

$$2c_1 = -c b_5 + \frac{c t_2^2}{2} (b_1 + b_3 + 2 b_4)$$

$$2c_2 = -c b_6 + \frac{c t_2^2}{2} (b_2 + b_3 + 2 b_4)$$

$$2c_3 = \frac{t_1^2}{2} (b_1 + b_3 + 2 b_4) - \frac{t_2^2}{2} (b_1 + b_3 + 2 b_4)$$

$$2c_4 = \frac{t_1^2}{2} (b_2 + b_3 + 2 b_4) - \frac{t_2^2}{2} (b_2 + b_3 + 2 b_4)$$

$$c_0 = \frac{c}{2} (b_5 + b_6) + \frac{t_1^3}{8} (b_1 + b_2 + 2 b_3 + 4 b_4)$$

$$+ \frac{t_2^3}{8} (b_1 + b_2 + 2 b_3 + 4 b_4) + \frac{t_1^3}{24} (b_1 + b_2 + 2 b_3 + 4 b_4)$$

$$+ \frac{t_2^3}{24} (b_1 + b_2 + 2 b_3 + 4 b_4).$$

(6.28)

It is important to note that in terms which include t_1, values of b_i appropriate to the upper face should be used; in terms which include t_2, values of b_i appropriate to the lower face should be used. The last two terms in the expression for c_0 are associated with local bending of the faces. Terms involving b_5 and b_6 are associated with shear deformation of the core.

The expression for the total energy of the system (equation (6.19)) may now be written:

$$U+V = \sum_{m=1}^{\infty} \sum_{n=1}^{\infty} \left\{ a_{mn}^2 f_{\min}(\lambda_{mn} \ldots) + \frac{N_x}{2} a_{mn}^2 (i_5)_{mn} \right.$$
$$\left. + \frac{N_y}{2} a_{mn}^2 (i_6)_{mn} - q a_{mn}(i_7)_{mn} \right\} \quad (6.29)$$

where $f_{\min}(\ )$ has been defined by equation (6.27). For equilibrium it is also necessary that $(U+V)$ should be stationary with respect to each a_{mn}:

$$\frac{\partial}{\partial a_{mn}}(U+V) = 2a_{mn}f_{\min} + N_x a_{mn}(i_5)_{mn}$$
$$+ N_y a_{mn}(i_6)_{mn} - q(i_7)_{mn} = 0 \quad (6.30a)$$

or
$$a_{mn} = \frac{q(i_7)_{mn}}{2f_{\min} + N_x(i_5)_{mn} + N_y(i_6)_{mn}}. \quad (6.30b)$$

In this way the amplitudes of the different modes of deformation may be evaluated. The stresses in the faces and in the core may be found by suitable differentiations of equation 6.17 in a manner analogous with that of Section 5.4.

If N_y is zero, the critical (buckling) value of N_x occurs when $a_{mn} \to \infty$ in equation (6.30b). Thus

$$(-N_x)_{\text{cr}} = P_{xmn} = \frac{2f_{\min}}{(i_5)_{mn}}. \quad (6.31)$$

6.6. Procedure in Particular Cases

The equations derived in the previous section may be applied to most kinds of sandwich in which the core makes a negligible

BENDING AND BUCKLING OF ORTHOTROPIC SANDWICH PANELS 115

contribution to the flexural rigidity of the sandwich. They can be used without modification for both bending and buckling problems and for different boundary conditions provided suitable orthogonal modes can be found. The procedure in any particular case is as follows:

(i) Determine suitable orthogonal modes $\phi_m(x)$, $\psi_n(y)$ which satisfy the boundary conditions, at least approximately.
(ii) Evaluate the integrals $(i_1)_{mn} - (i_7)_{mn}$ in equations (6.18).
(iii) Evaluate the quantities $(b_1)_{mn} - (b_6)_{mn}$ in equations (6.20). There are two sets of values $(b_1)_{mn} - (b_4)_{mn}$, one for each face.
(iv) Evaluate the quantities c_0, c_i, c_{ij}, c_{ii} in equations (6.28). There is only one set of these quantities for each mode.
(v) Evaluate the determinants in equation (6.26b) and the quantities x_1, x_2, x_3, x_4 ($= \lambda_{mn}$, μ_{mn}, $\lambda_{mn} q_{mn}$, $\mu_{mn} r_{mn}$) in equation (6.26a).
(vi) Evaluate $f_{\min}$ from equation (6.27) for each mode m, n.
(vii) The amplitude of the (m, n)th mode due to a uniform transverse pressure q and an edge load N_x (tension, per unit length) is given by equation (6.30b).
(viii) The critical value of the edge load N in the (m, n)th mode may be obtained from equation (6.31).

It is practicable to perform these calculations by hand if it is desired to determine either the amplitude of the first mode only ($m = n = 1$) or the critical edge load corresponding to one chosen mode. If, on the other hand, it is desired to obtain accurate results for deflections and stresses by taking the sum of the deflections and stresses in a number of different modes, or if it is desired to find the minimum buckling load by comparing the buckling loads corresponding to several different modes, it will almost certainly be necessary to rely on the services of a digital computer. The method described here has been used in the preparation of a series of graphs which define the critical value of N_x for a wide range of sandwich panels with different boundary conditions, face and core materials. [(1.17)]

It is sometimes possible to achieve considerable simplification in special cases, as in the next section.

6.7. Simply-supported Sandwich Plate with Identical Isotropic Thick Faces and "Isotropic" Core

The core is assumed to be isotropic in the sense that $G_{yz} = G_{zx} = G$, but the contribution of the core to the flexural rigidity of the sandwich is neglected. As usual, this is taken to imply that shear strains are constant through the thickness of the core. In the faces, which are identical,

$$E_x = E_y = E; \quad G_{xy} = E/2(1+\nu); \quad g = 1 - \nu^2.$$

The following modes are suitable for use with a simply-supported plate:

$$\phi_m(x) = \sin\frac{m\pi x}{a}, \quad \psi_n(y) = \sin\frac{n\pi y}{b}.$$

Because the faces are identical, the sandwich is symmetrical about the middle plane and $q = r = c/2$. It can also be shown that $\mu = \lambda$; this will not be proved here, but the result will be used to simplify the subsequent analysis. This simplification is only possible when the faces and the core are "isotropic".

From equation (6.20), using the results in Table 6.1,

$$\begin{aligned}
(b_1)_{mn} &= \frac{E}{g}\pi^4\frac{m^4}{a^4}\frac{ab}{4}, & (b_2)_{mn} &= \frac{E}{g}\pi^4\frac{n^4}{b^4}\frac{ab}{4}, \\
(b_3)_{mn} &= \frac{E\nu}{g}\pi^4\frac{m^2n^2}{a^2b^2}\frac{ab}{4}, & (b_4)_{mn} &= \frac{E(1-\nu)}{2g}\pi^4\frac{m^2n^2}{a^2b^2}\frac{ab}{4}, \\
(b_5)_{mn} &= G\pi^2\frac{m^2}{a^2}\frac{ab}{4}, & (b_6)_{mn} &= G\pi^2\frac{n^2}{b^2}\frac{ab}{4}.
\end{aligned} \quad (6.32)$$

(b_1–b_4 are the same for the upper and lower faces).

Equation (6.21) reduces to the following form:
$$f(\lambda_{mn}) = c_{11}\lambda_{mn}^2 + 2c_1\lambda_{mn} + c_0, \qquad (6.33a)$$
where
$$\left.\begin{aligned} c_{11} &= \frac{c}{2}(b_5+b_6) + \frac{c^2 t}{4}(b_1+b_2+2b_3+4b_4), \\ c_1 &= -\frac{c}{2}(b_5+b_6) + \frac{c^2 t}{4}(b_1+b_2+2b_3+4b_4), \\ c_0 &= \frac{c}{2}(b_5+b_6) + \frac{t^3}{3}(b_1+b_2+2b_3+4b_4) \end{aligned}\right\} \qquad (6.33b)$$

and
$$\left.\begin{aligned} b_5+b_6 &= \frac{\pi^2}{b^2}\frac{Gab}{4}\Omega, \\ b_1+b_2+2b_3+4b_4 &= \frac{\pi^4}{b^4}\frac{E}{g}\frac{ab}{4}\Omega^2, \end{aligned}\right\} \qquad (6.33c)$$

$$\Omega = m^2\frac{b^2}{a^2} + n^2. \qquad (6.33d)$$

In equations (6.33) the suffixes m, n have been omitted from c_0, c_1, c_{11} and b_1–b_6. The minimum value of $f(\lambda_{mn})$ may be found by differentiating (6.33a) with respect to λ_{mn} and equating the result to zero:

$$f(\lambda_{mn})\,(\text{minimum}) = \frac{\pi^4 E d^2 t a}{16 g b^3}\left(\frac{1}{1+\varrho\Omega} + \frac{t^2}{3d^2}\right), \qquad (6.34a)$$
where
$$\varrho = \frac{\pi^2}{b^2}\frac{E}{G}\frac{ct}{2g}. \qquad (6.34b)$$

The quantity ϱ is equal to π^2/b^2 times the ratio of the flexural rigidity $Ed^2t/2g$ (neglecting the local bending stiffness of the faces) to the shear stiffness Gd^2/c.

The minimum value of $f(\lambda_{mn})$ occurs when
$$\lambda_{mn} = \frac{1 - \dfrac{t}{c}\varrho\Omega}{1+\varrho\Omega}. \qquad (6.35)$$

The critical value of the edge load in the x-direction may be found from equation (6.31) with the following result:

$$P_{xmn} = \frac{\pi^2}{b^2} D_2 K_2, \quad (6.36a)$$

where

$$D_2 = \frac{Etd^2}{2g}, \quad (6.36b)$$

$$K_2 = \left(\frac{mb}{a} + \frac{n^2 a}{mb}\right)^2 \left\{\frac{1}{1 + \varrho\left(\dfrac{m^2 b^2}{a^2} + n^2\right)} + \frac{t^2}{3\,d^2}\right\}. \quad (6.36c)$$

$$m = 1, 2, 3, \ldots, \quad n = 1, 2, 3, \ldots$$

As usual, the smallest critical load is obtained by equating n to unity. Equation (6.36) differs from the corresponding equation for very thin faces (5.29) in the presence of the term $t^2/3d^2$ to account for the flexural rigidity of the faces and in the substitution of c for d in the expression for ϱ (6.34b); this latter change allows for the thickness of the faces in the geometry of the deformations. Figure 6.3 shows the variation of the minimum value of K_2 with a/b for various values of ϱ and for $t/d = 0$ and 0.3. These values of t/d represent the extreme values of any practical sandwich and it is clear that the influence of the face thickness is significant only for very weak cores ($\varrho > 0.4$, say) or for short plates ($a/b < 0.5$, say). A method for calculating K_2 in these circumstances is given in Chapter 10.

When t/d is equal to 0.3, all points to the left of the curve AA represent buckling into one half-wavelength ($m = 1$). A similar curve could be drawn for the other values of t/d, but as t/d is reduced the bulge extends further to the left. The extreme points AA remain at $a/b = \sqrt{2}$, however, because the transition from $m = 1$ to $m = 2$ must occur at that value in the extreme cases $\varrho = 0$ (no shear deformation of the core) and $\varrho = \infty$ (faces acting as independent plates).

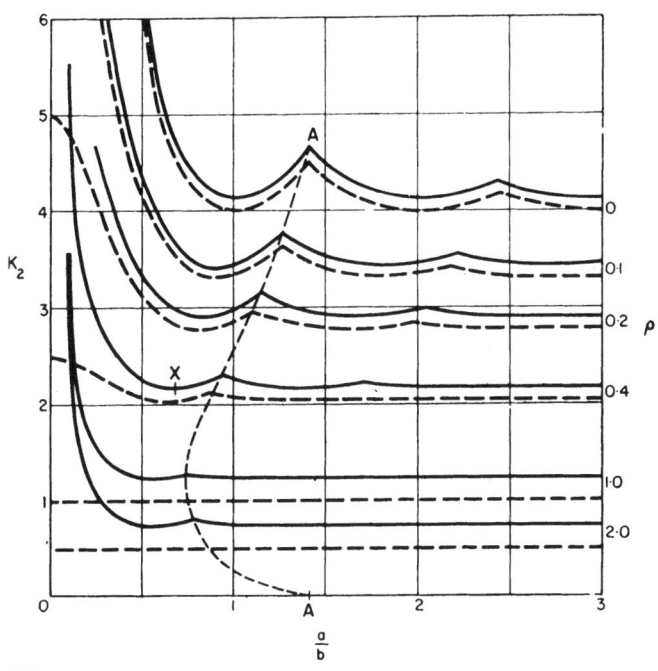

FIG. 6.3. Buckling coefficient K_2 in equation (6.36). Simply-supported isotropic sandwich panel with thick faces and uniform edge load in the x-direction. Broken and full lines represent the extreme cases $t/d = 0$, 0.3, respectively.

$$\left\{ \varrho = \frac{\pi^2}{2(1-v_f^2)} \frac{E_f}{G_c} \frac{tc}{b^2} \right\}.$$

The amplitude of the (m, n)th mode due to a uniform transverse pressure q in the absence of edge load may be found from equation (6.30b) by equating N_x and N_y to zero.

$$a_{mn} = \frac{16qb^4}{\pi^6 mn D_2} \frac{\zeta}{\Omega^2} \qquad m = 1, 3, 5, \ldots, \; n = 1, 3, 5, \ldots, \tag{6.37a}$$

where

$$\frac{1}{\zeta} = \left[\frac{1}{1+\varrho\Omega} + \frac{t^2}{3d^2} \right]. \tag{6.37b}$$

When the faces are *very* thin, c and d are approximately equal. The shear stiffness is then Gd and the value of ϱ given here becomes identical with the value given by equation (5.25c). Furthermore, t/d can be neglected in equation (6.37b); equations (6.37a) and (5.32) are then identical. Finally, when n is unity, equations (6.36) and (5.29) are identical.

Equation (6.37) may be used to evaluate the maximum deflection and stresses in the panel due to a uniform transverse pressure q. The strains for the various modes are obtained by substitution of the sinusoidal displacement functions of amplitude a_{mn} in the expressions (6.4), (6.6) and (6.8), noting that λ_{mn} is defined by equation (6.35), $\mu_{mn} = \lambda_{mn}$ and $q = r = c/2$. The stresses in each mode may be obtained from the strains in the usual way. The total deflections and stresses are obtained by summing the corresponding deflections and stresses in each mode (m, n).

The peak values of the deflections and stresses obtained in this way are listed below.

Deflection

$$w = \frac{qb^4}{D_2}\alpha_1. \qquad (6.38a)$$

Membrane stresses in lower face

$$\sigma_x = \frac{qb^2}{dt}(\alpha_3 + \nu\alpha_4); \qquad \sigma_y = \frac{qb^2}{dt}(\alpha_4 + \nu\alpha_3); \quad (6.38b, c)$$

$$\tau_{xy} = -\frac{qb^2}{dt}(1-\nu)\alpha_5. \qquad (6.38d)$$

Core shear stresses

$$\tau_{zx} = \frac{qb}{d}\alpha_6; \qquad \tau_{yz} = \frac{qb}{d}\alpha_7. \qquad (6.38e, f)$$

Local bending stresses in lower surface of lower face

$$\sigma_x = \frac{3qb^2}{t^2}(\alpha_8 + \nu\alpha_9); \qquad \sigma_y = \frac{3qb^2}{t^2}(\alpha_9 + \nu\alpha_8); \qquad (6.38\text{g, h})$$

$$\tau_{xy} = -\frac{3qb^2}{t^2}(1-\nu)\alpha_{10}. \qquad (6.38\text{i})$$

The dimensionless coefficients α_1 and α_3–α_{10} are listed in Table 6.2. They are functions of a/b (the aspect ratio of the plate), t/d and ϱ (a measure of the ratio of the bending stiffness to the shear stiffness).

TABLE 6.2

$$\alpha_1 = \frac{16}{\pi^6} \sum \sum \frac{\zeta}{mn\Omega^2}(-1)^{(m-1)/2}(-1)^{(n-1)/2},$$

$$\alpha_3 = \frac{16}{\pi^4} \sum \sum \frac{\zeta}{\Omega^2(1+\varrho\Omega)} \frac{m}{n} \frac{b^2}{a^2}(-1)^{(m-1)/2}(-1)^{(n-1)/2},$$

$$\alpha_4 = \frac{16}{\pi^4} \sum \sum \frac{\zeta}{\Omega^2(1+\varrho\Omega)} \frac{n}{m}(-1)^{(m-1)/2}(-1)^{(n-1)/2},$$

$$\alpha_5 = \frac{16}{\pi^4} \sum \sum \frac{\zeta}{\Omega^2(1+\varrho\Omega)} \frac{b}{a},$$

$$\alpha_6 = \frac{16}{\pi^3} \sum \sum \frac{\zeta}{\Omega(1+\varrho\Omega)} \frac{1}{n} \cdot \frac{b}{a}(-1)^{(n-1)/2},$$

$$\alpha_7 = \frac{16}{\pi^3} \sum \sum \frac{\zeta}{\Omega(1+\varrho\Omega)} \frac{1}{m}(-1)^{(m-1)/2},$$

$$\alpha_8 = \frac{16}{\pi^4} \cdot \frac{t^2}{3d^2} \sum \sum \frac{\zeta}{\Omega^2} \frac{m}{n} \frac{b^2}{a^2}(-1)^{(m-1)/2}(-1)^{(n-1)/2},$$

$$\alpha_9 = \frac{16}{\pi^4} \cdot \frac{t^2}{3d^2} \sum \sum \frac{\zeta}{\Omega^2} \frac{n}{m}(-1)^{(m-1)/2}(-1)^{(n-1)/2},$$

$$\alpha_{10} = \frac{16}{\pi^4} \frac{t^2}{3d^2} \sum \sum \frac{\zeta}{\Omega^2} \frac{b}{a}.$$

All summations are for $m = 1, 3, 5 \ldots$, $n = 1, 3, 5 \ldots$.

$$\varrho = \frac{\pi^2}{2(1-\nu^2)} \frac{E}{G} \frac{tc}{b^2}; \qquad \Omega = \frac{m^2b^2}{a^2}+n^2; \qquad \zeta = \left[(1+\varrho\Omega)^{-1}+\frac{t^2}{3d^2}\right]^{-1}.$$

Values of α_1 and α_3–α_{10} for a range of values of t/d and ϱ in a square plate are displayed in Fig. 6.4(a)–(c) (see pages 122–124).

The stresses and deflections in a long rectangular plate at points well away from the short edges may be obtained by treating the plate as a wide beam. In that case the results of Chapter 2 may be used, provided the modulus of elasticity of the face is everywhere divided by $(1 - \nu^2)$.

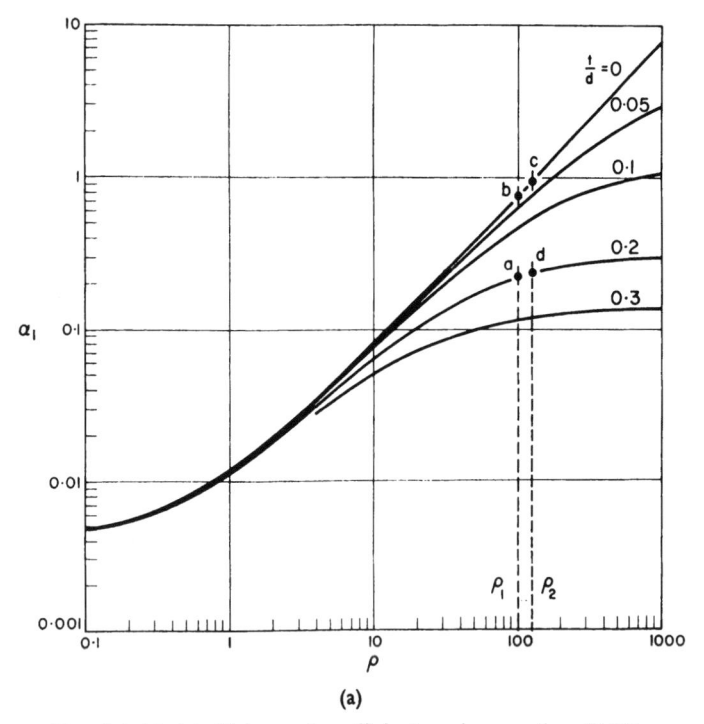

Fig. 6.4 (a)–(c). Values of coefficients α in equation (6.38). Isotropic panel with thick faces. (The summations in Table 6.2 are taken up to and including $m = n = 23$.)

It is not difficult to demonstrate that equations (5.35), (5.39) and (5.40), derived for an isotropic panel with very thin faces, represent a special case of equations (6.38) in which $c \to d$. The coefficients α_3–α_7 correspond to β_3–β_7.

(b)

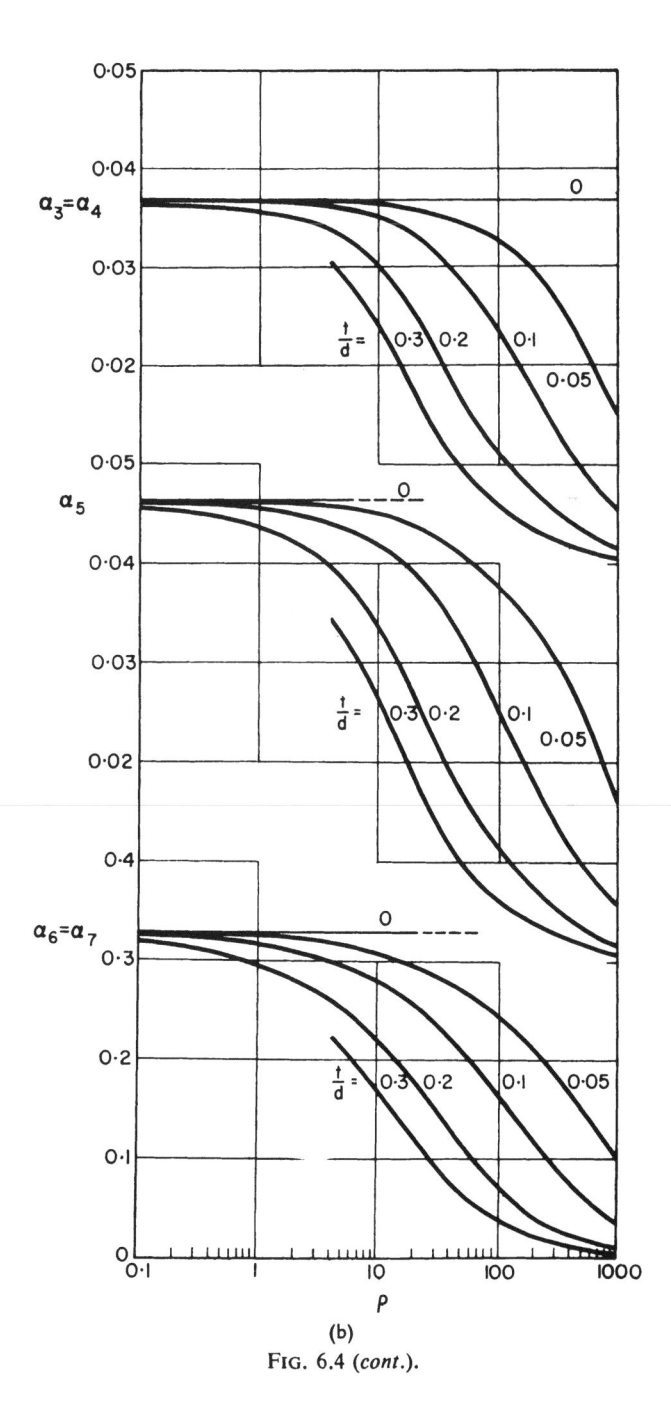

FIG. 6.4 (*cont.*).

Figure 6.4 may be used to provide a quick estimate of the importance of face thickness in any particular problem. The incorporation of a finite face thickness in the analysis alters the geometry of the deformation of the system (because plane sections no longer

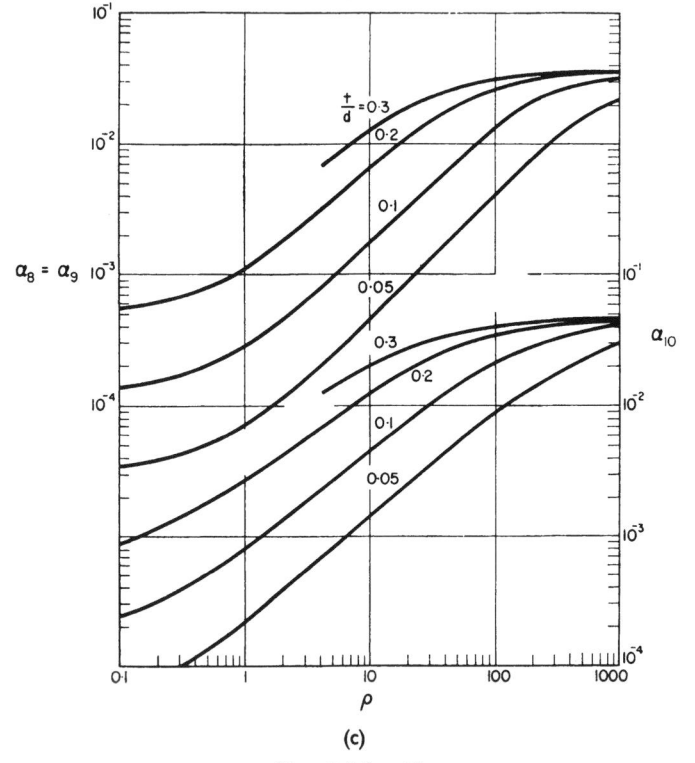

[Figure 6.4 (cont.): Log-log plot with ρ on the horizontal axis (0.1 to 1000) and $\alpha_8 = \alpha_9$ on the left vertical axis (10^{-4} to 10^{-1}), α_{10} on the right vertical axis (10^1 to 10^4). Curves are labeled with $\frac{t}{d} = 0.3, 0.2, 0.1, 0.05$.]

(c)

FIG. 6.4 (cont.).

remain plane but zigzag through the thicknesses of the faces and core); it also introduces the local bending stiffnesses of the faces. Consider, for example, the calculation of α_1 for a sandwich for which the correct value of ϱ is ϱ_1 (equation (6.34b)). The correct value of α_1 is given by point (a) in Fig. 6.4. If the bending stiffnesses of the faces are neglected, but the geometrical effect is retained, the value of α_1 is given by point (b) corresponding to $t/d = 0$. If, on the other hand, the geometrical effect is neglected (as in Chapter 5) so that $c \to d$, the shear stiffness Gd^2/c is incorrectly represented by the lower value Gd and a spurious value, ϱ_2, is obtained from equation (5.25c). The result is the point (c) if the face stiffness effect is also neglected, or the point (d) if it is not.

The points (b), (c) and (d) therefore represent varying degrees of approximation to the correct point (a).

CHAPTER 7

BENDING AND BUCKLING OF ORTHOTROPIC SANDWICH PANELS WITH THIN FACES; ALTERNATIVE SOLUTION BASED ON DIFFERENTIAL EQUATIONS OF SANDWICH PANEL

7.1. Introduction

In the previous chapter the behaviour of sandwich plates was examined by a detailed study of the strains in the core and the faces, and of the strain energy associated with those strains. The technique was developed by various workers at the U.S. Forest Products Laboratory as an extension of earlier studies at the Royal Aircraft Establishment. An alternative approach was adopted by Libove and Batdorf[2.1] and later developed by Seide and Stowell,[2.2] Robinson,[5.7] Harris and Auelmann[4.6] and others. In this alternative approach the character of the sandwich plate is defined by certain bending, twisting and shearing stiffnesses in the x- and y-directions. Differential equations may be set up in terms of these stiffnesses, or a strain energy function may be derived from them. The solution of the differential equations by the substitution of appropriate deflection functions is fairly straightforward in the simply-supported case.

In the work which follows, an outline of the method will be

BENDING AND BUCKLING OF ORTHOTROPIC SANDWICH PANELS

given. Its relation to the method of Chapter 6 will be discussed (see Sections 7.4, 7.5 and 9.2) and it will be used for the analysis of the bending and buckling of simply-supported orthotropic sandwich panels (including panels with corrugated cores).

7.2. Notation, Assumptions and Basic Equations

The analysis is applicable to a sandwich in which the faces are of different orthotropic materials and of different thicknesses. In the original form presented by Libove and Batdorf[2.1] it was limited to *very thin* faces, but the form to be described below is appropriate to *thin* faces (in which account is taken of the effect of face thickness of the geometry of the deformation). The principal axes of the face materials must be parallel with the sides of the panel. No allowance is made for the local bending stiffnesses of the faces about their own centroidal axes.

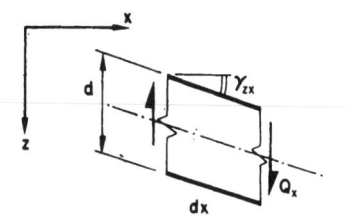

FIG. 7.1. Shear strain in zx-plane; very thin faces.

The core may be orthotropic and it is assumed to be antiplane and therefore of negligible stiffness in the xy-plane. Consequently the core makes no contribution to the bending stiffness of the sandwich† and the shear stresses are constant through the depth of the core. The core is also assumed to be infinitely stiff in the

---

† An exception to this is the corrugated-core sandwich, in which the core contributes significantly to the flexural rigidity in one direction. However, it is possible to apply the analysis to such sandwiches without much modification. See Section 7.6 (iv).

z-direction; any effects due to flattening or squashing of the core are therefore excluded.

The various forces which act on a small element of sandwich, of unit dimensions in the x-and y-directions, are shown in Fig. 5.2. The sign convention is identical with that used by Libove and Batdorf[2.1] and by Robinson.[5.7] It is also used by Timoshenko [35.3, 35.14] in his books on the behaviour of homogeneous plates, with which the following analysis may be compared.

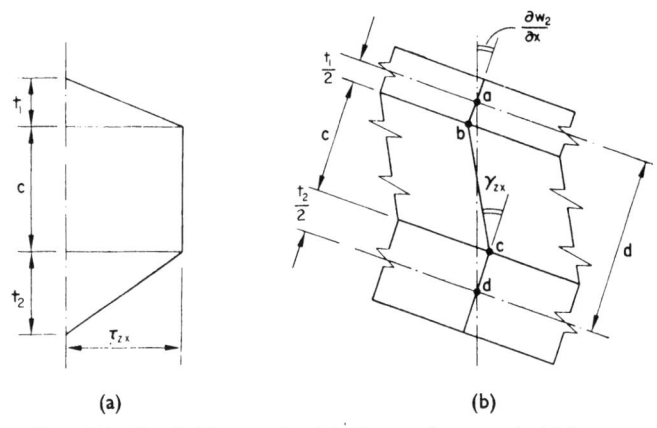

FIG. 7.2. Sandwich panel with faces of unequal thickness. (a) Distribution of the shear stress τ_{zx}. (b) Shear deformation in zx-plane.

The transverse deformation w of the plate may be regarded as the sum of two independent deformations w_1 and w_2. The first of these represents the bending deformation of the plate in the absence of shear strain in the core. The second component, w_2, represents the additional deformation associated with shear strain in the core. When the plate takes part in the deformation w_2, points in the middle planes of the faces (such as a and d in Fig. 7.2b) move in the z-direction only; consequently there are no face membrane forces associated with w_2.

Suppose now that a bending deformation w_1 occurs as the result of the application of a bending moment M_x alone. The resulting curvatures $\partial^2 w_1/\partial x^2$, $\partial^2 w_1/\partial y^2$ in the zx- and yz-planes are related to M_x in the following way:

$$\frac{\partial^2 w_1}{\partial x^2} = -\frac{M_x}{D_x}; \quad \frac{\partial^2 w_1}{\partial y^2} = -\nu_x \frac{\partial^2 w_1}{\partial x^2}. \qquad (7.1\text{a, b})$$

Here D_x is the flexural rigidity of the sandwich as a whole in the zx-plane; ν_x is in the nature of a Poisson's ratio. Formulae for D_x are given in Section 7.6.

Similarly the curvatures which are a consequence of the application of a bending moment M_y are:

$$\frac{\partial^2 w_1}{\partial y^2} = -\frac{M_y}{D_y}; \quad \frac{\partial^2 w_1}{\partial x^2} = -\nu_y \frac{\partial^2 w_1}{\partial y^2}. \qquad (7.2\text{a, b})$$

The corresponding relationship between the rate of twist and the twisting couple M_{xy} is:

$$\frac{\partial^2 w_1}{\partial x \, \partial y} = \frac{M_{xy}}{D_{xy}}. \qquad (7.3)$$

Equilibrium demands that

$$M_{yx} = -M_{xy}. \qquad (7.4)$$

The "Poisson's ratios" ν_x, ν_y are connected by the relationship

$$D_x \nu_y = D_y \nu_x. \qquad (7.5)$$

This may easily be demonstrated by the application of the reciprocal theorem to the two bending moments M_x and M_y and to the curvatures which they induce.

The transverse deformation w_2 is associated with shear strain in the core and therefore with the shear forces Q_x and Q_y. The relationship between w_2 and the shear forces may be defined in terms of core shear stiffnesses, D_{Qx} and D_{Qy}, as follows:†

$$Q_x = D_{Qx}\frac{\partial w_2}{\partial x}; \qquad Q_y = D_{Qy}\frac{\partial w_2}{\partial y}. \qquad (7.6a, b)$$

From equations (7.6) may be derived the curvatures and rate of twist due to the deformation w_2:

$$\frac{\partial^2 w_2}{\partial x^2} = \frac{1}{D_{Qx}}\frac{\partial Q_x}{\partial x}; \qquad \frac{\partial^2 w_2}{\partial y^2} = \frac{1}{D_{Qy}}\frac{\partial Q_y}{\partial y}; \qquad (7.7a, b)$$

$$\frac{\partial^2 w_2}{\partial x\, \partial y} = \frac{1}{2D_{Qx}}\frac{\partial Q_x}{\partial y} + \frac{1}{2D_{Qy}}\frac{\partial Q_y}{\partial x}. \qquad (7.7c)$$

The total curvatures and rate of twist of the plate are obtained by superimposing the separate contributions in equations (7.1), (7.2), (7.3) and (7.7):

$$\frac{\partial^2 w}{\partial x^2} = \frac{\partial^2 w_1}{\partial x^2} + \frac{\partial^2 w_2}{\partial x^2} = -\frac{M_x}{D_x} + \nu_y \frac{M_y}{D_y} + \frac{1}{D_{Qx}}\frac{\partial Q_x}{\partial x}, \qquad (7.8)$$

$$\frac{\partial^2 w}{\partial y^2} = -\frac{M_y}{D_y} + \nu_x \frac{M_x}{D_x} + \frac{1}{D_{Qy}}\frac{\partial Q_y}{\partial y}, \qquad (7.9)$$

$$\frac{\partial^2 w}{\partial x\, \partial y} = \frac{M_{xy}}{D_{xy}} + \frac{1}{2D_{Qx}}\frac{\partial Q_x}{\partial y} + \frac{1}{2D_{Qy}}\frac{\partial Q_y}{\partial x}. \qquad (7.10)$$

These are the fundamental equations which define the stiffness of the plate in bending and shear.

The equations which describe the equilibrium of a small element of the plate are the same as those for homogeneous plates. For example, equations (7.11)–(7.13) express the conditions for equilibrium in the x-, y- and z-directions, respectively:

$$\frac{\partial N_x}{\partial x} + \frac{\partial N_{xy}}{\partial y} = 0, \qquad (7.11)$$

† It is assumed that there is no relative rigid-body movement of the faces in the xy-plane. This is satisfactory if the load is symmetrical or if there are edge stiffeners.

$$\frac{\partial N_y}{\partial y} + \frac{\partial N_{xy}}{\partial x} = 0, \qquad (7.12)$$

$$\frac{\partial Q_x}{\partial x} + \frac{\partial Q_y}{\partial y} + q(x,y) + N_x \frac{\partial^2 w}{\partial x^2} + N_y \frac{\partial^2 w}{\partial y^2} + 2N_{xy} \frac{\partial^2 w}{\partial x \, \partial y} = 0. \qquad (7.13)$$

Equations (7.14)–(7.16) express the conditions for equilibrium of a plate element about the y-, x- and z-axes, respectively:

$$\frac{\partial M_x}{\partial x} - \frac{\partial M_{xy}}{\partial y} - Q_x = 0, \qquad (7.14)$$

$$\frac{\partial M_y}{\partial y} - \frac{\partial M_{xy}}{\partial x} - Q_y = 0, \qquad (7.15)$$

$$N_{yx} = N_{xy}. \qquad (7.16)$$

Derivations of these equations may be found in standard works of reference, such as Timoshenko[35.3] (Chapter 4).

7.3. General Differential Equations

It will be assumed that the forces N are constant throughout the plate, so that equations (7.11) and (7.12) are automatically satisfied and need not be considered further. This implies that the distribution of the forces N does not change as the plate bends; the analysis is therefore restricted to small deflections.

It is convenient to invert the equations (7.8)–(7.10):

$$M_x = -\frac{D_x}{g} \left\{ \frac{\partial}{\partial x} \left(\frac{\partial w}{\partial x} - \frac{Q_x}{D_{Qx}} \right) + v_y \frac{\partial}{\partial y} \left(\frac{\partial w}{\partial y} - \frac{Q_y}{D_{Qy}} \right) \right\}, \quad (7.17)$$

$$M_y = -\frac{D_y}{g} \left\{ \frac{\partial}{\partial y} \left(\frac{\partial w}{\partial y} - \frac{Q_y}{D_{Qy}} \right) + v_x \frac{\partial}{\partial x} \left(\frac{\partial w}{\partial x} - \frac{Q_x}{D_{Qx}} \right) \right\}, \quad (7.18)$$

$$M_{xy} = \frac{D_{xy}}{2} \left\{ \frac{\partial}{\partial x} \left(\frac{\partial w}{\partial y} - \frac{Q_y}{D_{Qy}} \right) + \frac{\partial}{\partial y} \left(\frac{\partial w}{\partial x} - \frac{Q_x}{D_{Qx}} \right) \right\}, \qquad (7.19)$$

where $g = 1 - v_x v_y$.

These bending and twisting moments may be substituted in equations (7.14) and (7.15) with the following results:

$$\left\{-D_{xy}\frac{\partial^3}{\partial x\,\partial y^2}-\frac{D_x}{g}\left(v_y\frac{\partial^3}{\partial x\,\partial y^2}+\frac{\partial^3}{\partial x^3}\right)\right\}w$$

$$+\left\{\frac{1}{2}\frac{D_{xy}}{D_{Qx}}\frac{\partial^2}{\partial y^2}+\frac{D_x}{gD_{Qx}}\frac{\partial^2}{\partial x^2}-1\right\}Q_x$$

$$+\left\{\frac{1}{2}\frac{D_{xy}}{D_{Qy}}\frac{\partial^2}{\partial x\,\partial y}+\frac{D_x v_y}{gD_{Qy}}\frac{\partial^2}{\partial x\,\partial y}\right\}Q_y = 0, \quad (7.20)$$

$$\left\{-D_{xy}\frac{\partial^3}{\partial x^2\,\partial y}-\frac{D_y}{g}\left(v_x\frac{\partial^3}{\partial x^2\,\partial y}+\frac{\partial^3}{\partial y^3}\right)\right\}w$$

$$+\left\{\frac{1}{2}\frac{D_{xy}}{D_{Qx}}\frac{\partial^2}{\partial x\,\partial y}+\frac{D_y v_x}{gD_{Qx}}\frac{\partial^2}{\partial x\,\partial y}\right\}Q_x$$

$$+\left\{\frac{1}{2}\frac{D_{xy}}{D_{Qy}}\frac{\partial^2}{\partial x^2}+\frac{D_y}{gD_{Qy}}\frac{\partial^2}{\partial y^2}-1\right\}Q_y = 0. \quad (7.21)$$

These, together with equation (7.13), are three equations containing only three unknowns Q_x, Q_y and w. They were arranged by Libove and Batdorf[2.1] in the form

$$[D]w = -[M]q, \quad (7.22a)$$
$$[D]Q_x = -[N]q, \quad (7.22b)$$
$$[D]Q_y = -[P]q, \quad (7.22c)$$

where $[D]$, $[M]$, $[N]$, $[P]$ are rather lengthy differential operators. Equations (7.22) might be solved, in theory, for each of the unknowns w, Q_x and Q_y in terms of a given load function $q(x, y)$ However, it is usually more convenient (at least for simply-supported panels) to assume the form of the solution and to insert it directly into equations (7.13), (7.20) and (7.21). This was essentially the procedure adopted by Robinson [5.7] and an equivalent procedure will be used in the following sections.

Libove and Batdorf [2.1] also derived expressions for the strain energy in the plate in terms of the derivatives of Q_x, Q_y and w (their equations 26, 27, 28).

7.4. Simplified Form of Differential Equations

Consider a point "a" on the middle plane of the upper face and a corresponding point "d" on the middle plane of the lower face such that the line ad is parallel with the z-axis when the plate is flat and unstressed.

When a transverse bending deformation (w_1) occurs, the line ad remains perpendicular to the middle planes of the upper and lower faces. The rotations of the line ad in the zx-and yz-planes are therefore $\partial w_1/\partial x$ and $\partial w_1/\partial y$, and these are equal to the slopes of the middle planes of the faces.

When a transverse shear deformation (w_2) occurs, the line ad remains vertical as in Fig. 7.2b, but the slopes of the middle planes of the faces are $\partial w_2/\partial x$ and $\partial w_2/\partial y$.

When both bending and shear deformations combine in a total transverse deformation ($w = w_1 + w_2$), the slopes of the middle planes of the faces are $\partial w/\partial x$ and $\partial w/\partial y$, whereas the rotations of the line ad are only $\partial w_1/\partial x$ and $\partial w_1/\partial y$. It is convenient to relate the slopes and the rotations by the following equations:

$$\frac{\partial w_1}{\partial x} = \lambda' \frac{\partial w}{\partial x}, \tag{7.23a}$$

$$\frac{\partial w_1}{\partial y} = \mu' \frac{\partial w}{\partial y}. \tag{7.23b}$$

The equations (7.6), which define the shear stiffnesses D_{Qx} and D_{Qy}, may be combined with equations (7.23) in the following manner:

$$\lambda' \frac{\partial w}{\partial x} = \frac{\partial w_1}{\partial x} = \frac{\partial w}{\partial x} - \frac{\partial w_2}{\partial x} = \frac{\partial w}{\partial x} - \frac{Q_x}{D_{Qx}}. \tag{7.24a}$$

Similarly,

$$\mu' \frac{\partial w}{\partial y} = \frac{\partial w}{\partial y} - \frac{Q_y}{D_{Qy}}. \tag{7.24b}$$

The same expressions occur on the right-hand sides of equations (7.24) and (7.17)–(7.19); the latter group of equations may therefore be simplified considerably by substitutions for

$$\left(\frac{\partial w}{\partial x} - \frac{Q_x}{D_{Qx}}\right) \quad \text{and} \quad \left(\frac{\partial w}{\partial y} - \frac{Q_y}{D_{Qy}}\right).$$

In general, λ' and μ' vary from point to point over the surface of the plate. It will be shown in Section 7.7, however, that a solution can be obtained for plates with simply supported edges under certain loads when λ' and μ' are constants. If, then, λ' and μ' are assumed to be constants, the equations (7.17)–(7.19) may be written as follows:

$$M_x = -\frac{D_x}{g}\left\{\lambda' \frac{\partial^2 w}{\partial x^2} + \mu' v_y \frac{\partial^2 w}{\partial y^2}\right\}, \tag{7.25}$$

$$M_y = -\frac{D_y}{g}\left\{\mu' \frac{\partial^2 w}{\partial y^2} + \lambda' v_x \frac{\partial^2 w}{\partial x^2}\right\}, \tag{7.26}$$

$$M_{xy} = \frac{D_{xy}}{2}(\lambda' + \mu') \frac{\partial^2 w}{\partial x \, \partial y}. \tag{7.27}$$

Substitution in equations (7.14), (7.15) and (7.13) for M_x, M_y and M_{xy} (from equations (7.25)–(7.27)) and for Q_x and Q_y (from equations (7.24)) provides three basic equations:

$$\frac{\partial}{\partial x}\left[\lambda'\left(\frac{D_x}{g}\frac{\partial^2 w}{\partial x^2} + \frac{D_{xy}}{2}\frac{\partial^2 w}{\partial y^2} - D_{Qx}w\right)\right.$$
$$\left. + \mu'\left(\frac{D_x v_y}{g} + \frac{D_{xy}}{2}\right)\frac{\partial^2 w}{\partial y^2} + D_{Qx}w\right] = 0, \tag{7.28}$$

$$\frac{\partial}{\partial y}\left[\mu'\left(\frac{D_y}{g}\frac{\partial^2 w}{\partial y^2} + \frac{D_{xy}}{2}\frac{\partial^2 w}{\partial x^2} - D_{Qy}w\right)\right.$$
$$\left. + \lambda'\left(\frac{D_y v_x}{g} + \frac{D_{xy}}{2}\right)\frac{\partial^2 w}{\partial x^2} + D_{Qy}w\right] = 0, \tag{7.29}$$

$$D_{Qx}(1-\lambda')\frac{\partial^2 w}{\partial x^2} + D_{Qy}(1-\mu')\frac{\partial^2 w}{\partial y^2} + q(x,y) + N_x\frac{\partial^2 w}{\partial x^2}$$
$$+ N_y\frac{\partial^2 w}{\partial y^2} + 2N_{xy}\frac{\partial^2 w}{\partial x \partial y} = 0. \quad (7.30)$$

The solution of these basic equations (7.28)–(7.30) for $w(x, y)$ and for the constants λ' μ' will be given in detail in Section 7.7.

7.5. Core Shear Strains and the Effect of Face Thickness

The vertical distribution of the shear stress τ_{zx} is illustrated in Fig. 7.2a, from which it follows that the shear force is

$$Q_x = \tau_{zx}\left(c + \frac{t_1+t_2}{2}\right) = \tau_{zx}d = G_{zx}\gamma_{zx}d, \quad (7.31)$$

where d is the distance between the centre-lines of the faces and γ_{zx} is the core shear strain. This equation may be used to determine the core shear stress or strain when the shear force is known.

In Fig. 7.2b is shown a section of the sandwich plate in the zx-plane undergoing a transverse shear displacement w_2. From the geometry of the deformation it is easy to verify that the shear strain is

$$\gamma_{zx} = \frac{d}{c}\frac{\partial w_2}{\partial x}. \quad (7.32)$$

The elimination of γ_{zx} between equations (7.31) and (7.32) leads to the following:

$$Q_x = \left(G_{zx}\frac{d^2}{c}\right)\frac{\partial w_2}{\partial x}. \quad (7.33)$$

If this result is compared with equation (7.6), it can be seen that the shear stiffness D_{Qx} (and the analogous shear stiffness D_{Qy}) are given by:

$$D_{Qx} = G_{zx}\frac{d^2}{c}; \quad D_{Qy} = G_{yz}\frac{d^2}{c}. \quad (7.34a, b)$$

This important result is valid for all cores which may be considered as approximately homogeneous.

The relationship between the analyses of Chapters 6 and 7 may be made more clear in the following way. From equations (7.6a) and (7.23a), the shear force Q_x may be written in terms of the coefficient λ':

$$Q_x = D_{Qx}\frac{\partial w_2}{\partial x} = D_{Qx}\left(\frac{\partial w}{\partial x} - \frac{\partial w_1}{\partial x}\right) = D_{Qx}(1-\lambda')\frac{\partial w}{\partial x}. \quad (7.35)$$

Elimination of Q_x and D_{Qx} by the use of equations (7.31) and (7.34a) leads to the following expression for the core shear strain:

$$\gamma_{zx} = \frac{d}{c}(1-\lambda')\frac{\partial w}{\partial x}. \quad (7.36a)$$

The analogous expression for the core shear strain in the yz-plane is

$$\gamma_{yz} = \frac{d}{c}(1-\mu')\frac{\partial w}{\partial y}. \quad (7.36b)$$

These expressions are directly comparable with equations (6.4) and as a consequence the relationship between λ and λ' may be expressed thus:

$$\lambda' = (t+c\lambda)/d; \quad \lambda = (d\lambda'-t)/c. \quad (7.37)$$

Corresponding expressions relate μ and μ'. In the extreme case of a core which is perfectly rigid in shear, λ and λ' are both unity. In the other extreme case, when the core has no shear stiffness at all, λ is equal to $-t/c$ but λ' vanishes.

The connection between λ' and μ' on the one hand and λ and μ on the other is now apparent. If the sandwich section in Fig. 7.2b is subjected to a total displacement w (not merely w_2 as shown) then λ represents the rotation of the line bc divided by $\partial w/\partial x$, whereas λ' represents the rotation of ad divided by $\partial w/\partial x$.

It is a matter of convenience whether λ' and μ' are used in preference to λ and μ. The former are used in this chapter and the latter are used in the previous chapters simply in order to keep the equations in the forms adopted in earlier literature. The distinction between λ and λ' and between μ and μ' vanishes when the faces are very thin.

7.6. Evaluation of the Stiffnesses D_x, D_{Qx}, etc.

Expressions for the various bending, twisting and shearing stiffnesses are given below without proof. The simplest cases are listed first.

(i) *Isotropic faces and core; faces of equal thickness and similar material*

$$D_x = D_y = E_f t\, d^2/2; \qquad D_{xy} = E_f t\, d^2/2(1+\nu_f);$$
$$D_{Qx} = D_{Qy} = G_c\, d^2/c.$$

(ii) *Orthotropic faces and core; faces of equal thickness and similar material*

$$D_x = E_x t\, d^2/2; \qquad D_y = E_y t\, d^2/2; \qquad D_{xy} = G_{xy} t\, d^2;$$
$$\nu_y E_x = \nu_x E_y;$$
$$D_{Qx} = G_{zx}\, d^2/c; \qquad D_{Qy} = G_{yz}\, d^2/c.$$

(G_{xy} refers to the faces: G_{zx} and G_{yz} to the core.)

(iii) *Orthotropic faces and core; faces of unequal thickness and different materials*

$$D_x = d^2\left(\frac{1}{E_{x1}t_1} + \frac{1}{E_{x2}t_2}\right)^{-1}; \qquad D_y = d^2\left(\frac{1}{E_{y1}t_1} + \frac{1}{E_{y2}t_2}\right)^{-1};$$
$$D_{xy} = 2\, d^2\left(\frac{1}{G_{xy1}t_1} + \frac{1}{G_{xy2}t_2}\right)^{-1}; \qquad \nu_y E_x = \nu_x E_y;$$
$$D_{Qx} = G_{zx}\, d^2/c; \qquad D_{Qy} = G_{yz}\, d^2/c.$$

(Suffixes 1 and 2 refer to the upper and lower faces respectively.)

(iv) *Corrugated-core sandwich with equal isotropic faces and a symmetrical core*

If the corrugations run in the x-direction, it is usual to make some allowance for the contribution of the core to the flexural rigidity of the sandwich in that direction. Then, even if the faces themselves are isotropic, it is to be expected that D_x will be greater than D_y. Because of this flexural rigidity of the core it cannot be assumed that the shear stresses in the zx-plane are constant through the depth of the core; some modification to the theory would therefore be required to predict the effect of core shear deformation in the zx-plane. Fortunately, the shear stiffness in the direction of the corrugations is usually so great that it can be taken as infinite. It is therefore necessary only to consider the effect of shear deformations in the plane perpendicular to the corrugations and the results of this chapter can be used as they stand.

Precise expressions for the stiffnesses of a corrugated-core sandwich are given by Libove and Hubka[12.4]; the results which follow are approximations:

$$D_x = E_f t \, d^2/2 + E_c I_c; \qquad D_y = \frac{E_f t \, d^2/2}{1 - \dfrac{v_f^2}{1 + (E_f t \, d^2)/(2 E_c I_c)}};$$

$$D_{xy} = E_f t \, d^2/2(1 + v_f),$$

$$D_{Qx} = \infty; \qquad D_{Qy} = \frac{S \, dE_c}{1 - v_c^2} \left(\frac{t_c}{d_c}\right)^3,$$

$$v_x = v_f; \qquad v_y = v_f D_y/D_x.$$

In these equations the suffixes f and c refer to the faces and core respectively. The dimensions t_c and d_c are defined in Fig. 7.3 and

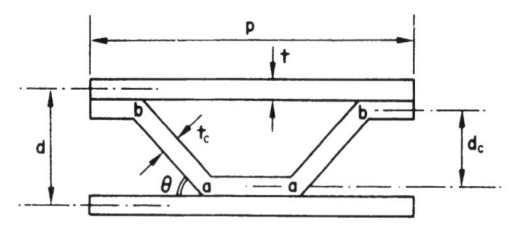

Fig. 7.3. Dimensions of sandwich with corrugated core.

I_c represents the second moment of area of the cross-section of the core per unit width in the y-direction. The coefficient S takes values in the range 0·5–15, depending on the geometry of the cross-section of the sandwich; reference should be made to the graphs prepared by Libove and Hubka[12.4] or to equivalent graphs in ref. (5.10).

When the contribution $(E_c\,I_c)$ of the core is small, D_y may be written simply as $E_f t\, d^2/2$.

7.7. Solution for Simply-supported Orthotropic Panel with Edge Loads and Sinusoidal Transverse Load

Consider a simply-supported orthotropic sandwich panel with edge loads N_x, N_y, N_{xy} and a sinusoidally distributed transverse load $q(x, y)$ where:

$$q(x, y) = q_{mn} \sin \alpha x \sin \beta y \qquad (7.38)$$

and

$$\alpha = \frac{m\pi}{a}, \quad \beta = \frac{n\pi}{b}, \quad m = 1, 2, 3, \ldots, \quad n = 1, 2, 3, \ldots$$

It will be assumed that a solution for the transverse displacements can be found in the form:

$$w = a_{mn} \sin \alpha x \sin \beta y. \qquad (7.39)$$

(It was demonstrated in Section 5.6 that a deflection function of this kind satisfies the boundary conditions for simply-supported sandwich panels with edge stiffeners which are rigid in their own planes.)

It will also be assumed that λ' and μ' are constants, independent of x and y, so that equations (7.28)–(7.30) are valid. Substitution for w and $q(x, y)$ in these equations gives the following result:

$$\lambda'\left(\frac{D_x}{g}\alpha^2 + \frac{D_{xy}}{2}\beta^2 + D_{Qx}\right) + \mu'\left(\frac{D_x v_y}{g} + \frac{D_{xy}}{2}\right)\beta^2 - D_{Qx} = 0,$$

$$(7.40)$$

$$\lambda'\left(\frac{D_y v_x}{g}+\frac{D_{xy}}{2}\right)\alpha^2+\mu'\left(\frac{D_y}{g}\beta^2+\frac{D_{xy}}{2}\alpha^2+D_{Qy}\right)-D_{Qy}=0,$$

(7.41)

$$\left\{D_{Qx}(1-\lambda')\alpha^2+D_{Qy}(1-\mu')\beta^2-\frac{q_{mn}}{a_{mn}}+N_x\alpha^2+N_y\beta^2\right\}\sin\alpha x\sin\beta y$$

$$-2N_{xy}\alpha\beta\cos\alpha x\cos\beta y=0 \quad \text{(provided } a_{mn}\neq 0\text{).} \quad (7.42)$$

Evidently equations (7.40) and (7.41) can be solved for λ' and μ' independently of the values assigned to N_x, N_y and N_{xy}. The resulting values of λ' and μ' can then be inserted in equation (7.42). However, it is clear from this last equation that a solution is only possible when $N_{xy}=0$ and $\sin\alpha x\sin\beta y$ cancels, leaving:

$$D_{Qx}(1-\lambda')\alpha^2+D_{Qy}(1-\mu')\beta^2-\frac{q_{mn}}{a_{mn}}+N_x\alpha^2+N_y\beta^2=0.$$

$$(a_{mn}\neq 0) \quad (7.43)$$

In other words, equation (7.39) satisfies the differential equations of the problem only in the absence of edge shearing forces N_{xy}. The solution of equations (7.40) and (7.41) may be written in the following form:

$$\lambda'=\frac{\delta_1}{\Psi}, \quad \mu'=\frac{\delta_2}{\Psi}, \quad (7.44a, b)$$

where δ_1, δ_2, Ψ are defined in Table 7.1. In that table two new quantities have been introduced:

$$B_{Qx}=b^2D_{Qx}, \quad B_{Qy}=b^2D_{Qy}. \quad (7.45a, b)$$

These have been used in preference to D_{Qx} and D_{Qy} because they have the same dimensions as D_x, D_y and D_{xy}. All the quantities λ', μ', Ψ, δ_1, δ_2 are non-dimensional.

It will later be found convenient to have available expressions for $B_{Qx}(1-\lambda')$ and $B_{Qy}(1-\mu')$:

$$B_{Qx}(1-\lambda')=\frac{\delta_3}{\Psi}, \quad B_{Qy}(1-\mu')=\frac{\delta_4}{\Psi}. \quad (7.46a, b)$$

BENDING AND BUCKLING OF ORTHOTROPIC SANDWICH PANELS

TABLE 7.1

$$\delta_1 = \frac{D_{xy}}{2B_{Qy}}\pi^2 \frac{m^2 b^2}{a^2} - \left\{\frac{D_x v_y}{B_{Qx}} - \frac{D_y}{B_{Qy}} + \frac{g}{2}\frac{D_{xy}}{B_{Qx}}\right\}\frac{n^2\pi^2}{g} + 1,$$

$$\delta_2 = -\left\{\frac{D_y v_x}{B_{Qy}} - \frac{D_x}{B_{Qx}} + \frac{g}{2}\frac{D_{xy}}{B_{Qy}}\right\}\frac{\pi^2 m^2}{g}\frac{b^2}{a^2} + \frac{D_{xy}}{2B_{Qx}}n^2\pi^2 + 1,$$

$$\delta_3 = \frac{\pi^4}{2gB_{Qy}}\left\{D_{xy}D_x\frac{m^4 b^4}{a^4} + D_{xy}D_y n^4 + \frac{2m^2 n^2 b^2}{a^2}(D_x D_y - D_{xy}D_x v_y)\right\}$$

$$+ \frac{m^2 b^2}{a^2}\frac{\pi^2}{g}D_x + \frac{n^2\pi^2}{g}(D_x v_y + g D_{xy}),$$

$$\delta_4 = \frac{\pi^4}{2gB_{Qx}}\left\{D_{xy}D_x\frac{m^4 b^4}{a^4} + D_{xy}D_y n^4 + 2m^2 n^2 \frac{b^2}{a^2}(D_x D_y - D_{xy}D_x v_y)\right\}$$

$$+ \frac{m^2 b^2}{a^2}\frac{\pi^2}{g}(D_y v_x + g D_{xy}) + n^2 \frac{\pi^2}{g}D_y,$$

$$\delta_5 = \frac{\pi^2}{2g}\left\{\frac{m^2}{B_{Qy}}\frac{b^2}{a^2} + \frac{n^2}{B_{Qx}}\right\}\left\{D_{xy}\frac{m^4 b^4}{a^4} + \frac{D_{xy}D_y}{D_x}n^4\right.$$

$$\left. + 2m^2 n^2 \frac{b^2}{a^2}(D_y - v_y D_{xy})\right\}$$

$$+ \frac{1}{g}\left\{\frac{m^4 b^4}{a^4} + \frac{D_y}{D_x}n^4 + 2m^2 n^2 \frac{b^2}{a^2}\left(v_y + \frac{D_{xy}g}{D_x}\right)\right\},$$

$$\Psi = \frac{\pi^4}{2gB_{Qx}B_{Qy}}\left\{D_{xy}D_x\frac{m^4 b^4}{a^4} + D_{xy}D_y n^4 + 2m^2 n^2 \frac{b^2}{a^2}(D_x D_y - D_{xy}D_x v_y)\right\}$$

$$+ \frac{\pi^2}{g}\frac{m^2 b^2}{a^2}\left\{\frac{D_x}{B_{Qx}} + \frac{g}{2}\frac{D_{xy}}{B_{Qy}}\right\} + \frac{\pi^2}{g}n^2\left\{\frac{D_y}{B_{Qy}} + \frac{g}{2}\frac{D_{xy}}{B_{Qx}}\right\} + 1.$$

The quantities δ_3, δ_4 are defined in Table 7.1.

If substitution is made for $D_{Qx}(1-\lambda')$ and $D_{Qy}(1-\mu')$ from equations (7.46) in equation (7.43), and the tensile forces N_x, N_y are replaced by $-P_x$, $-P_y$, the result is an equation which contains a_{mn} as the only unknown:

$$D_x \Phi_0 \frac{\pi^4}{b^4} - \frac{q_{mn}}{a_{mn}} - P_x \frac{m^2\pi^2}{a^2} - P_y \frac{n^2\pi^2}{b^2} = 0, \qquad (7.47a)$$

where

$$\Phi_0 = \frac{\delta_5}{\Psi} \qquad (7.47b)$$

and δ_5 is defined in Table 7.1.

A final re-arrangement of equation (7.47a) provides the desired expression for the amplitude of the transverse displacements.

$$a_{mn} = \frac{q_{mn}b^4}{\pi^2 D_x \Theta}, \quad (7.48a)$$

where

$$\Theta = \Phi_0 \pi^2 - \frac{P_x b^2}{D_x}\left(\frac{m^2 b^2}{a^2} + \frac{P_y}{P_x} n^2\right). \quad (7.48b)$$

The solution to the problem has therefore been found in equations (7.44) and (7.48), which define λ', μ' and a_{mn}.

The quantities λ', μ', $\delta_1 - \delta_5$, Ψ and Φ_0 are independent of the magnitudes of the loads and depend only on the stiffnesses (such as D_x, B_{Qx}), the Poisson's ratios, the aspect ratio of the panel (b/a) and the values of m and n. The equations (7.44)–(7.48) have been formulated in such a way that they remain valid in any of the extreme case $B_{Qx} \to \infty$, $B_{Qy} \to \infty$ or $D_{xy} \to 0$. They are not valid if B_{Qx} or B_{Qy} is zero. If such a situation is likely to occur it is advisable to factor the numerators and denominators in equations (7.44), (7.46) (7.47b) by $B_{Qx} B_{Qy}$.

The displacements are given by equation (7.39). If only odd values of m and n are being considered,[†] the peak value of w always occurs at the centre of the panel, $x = a/2$, $y = b/2$, in which case:

$$w_{\max} = a_{mn}(-1)^{(m-1)/2}(-1)^{(n-1)/2} = \frac{q_{mn}b^4}{\pi^2 D_x \Theta}(-1)^{(m-1)/2}(-1)^{(n-1)/2}. \quad (7.49)$$

The bending moments are given by equations (7.25), (7.26) and (7.39). The peak value for the odd modes occurs again at $x = a/2$, $y = b/2$. Equation (7.48) is used to eliminate a_{mn}.

$$(M_x)_{\max} = \frac{q_{mn}b^2}{g\Theta}\left\{\lambda' m^2 \frac{b^2}{a^2} + \mu' \nu_y n^2\right\}(-1)^{(m-1)/2}(-1)^{(n-1)/2}, \quad (7.50)$$

† Unless the loading is very unsymmetrical, the even modes are usually less important than the odd modes. See, for example, Section 7.9 in which the even modes are all zero.

$$(M_y)_{\max} = \frac{q_{mn}b^2}{g\Theta} \frac{D_y}{D_x} \left\{ \mu' n^2 + \lambda' \nu_x \frac{m^2 b^2}{a^2} \right\} (-1)^{(m-1)/2} (-1)^{(n-1)/2}. \tag{7.51}$$

The twisting moment is given by equations (7.27), (7.39) and (7.48), and the peak value for any mode occurs at $x = y = 0$:

$$(M_{xy})_{\max} = \frac{q_{mn}b^2}{2\Theta} \frac{D_{xy}}{D_x} (\lambda' + \mu') mn \frac{b}{a}. \tag{7.52}$$

The shear forces are defined by equation (7.35) and its analogue in the y-direction:

$$Q_x = D_{Qx}(1-\lambda') \frac{\partial w}{\partial x}; \qquad Q_y = D_{Qx}(1-\mu') \frac{\partial w}{\partial y}. \tag{7.53a, b}$$

The value of w is again taken from equation (7.39). For odd modes the peak value of Q_x occurs at $x = 0$, $y = b/2$ and of Q_y at $x = a/2$, $y = 0$. Hence:

$$(Q_x)_{\max} = \frac{q_{mn}b}{\pi\Theta} \frac{B_{Qx}(1-\lambda')}{D_x} m \frac{b}{a} (-1)^{(n-1)/2}, \tag{7.54}$$

$$(Q_y)_{\max} = \frac{q_{mn}b}{\pi\Theta} \frac{B_{Qy}(1-\mu')}{D_x} n(-1)^{(m-1)/2}. \tag{7.55}$$

Equations (7.49)–(7.52), (7.54) and (7.55) define the effects which are likely to be of greatest practical interest.

7.8. Buckling of Simply-supported Orthotropic Panel

In the preceding section the expression for the amplitude a_{mn} of the transverse displacements (7.48a) contained in the denominator the term:

$$\Theta = \Phi_0 \pi^2 - \frac{P_x b^2}{D_x} \left\{ \frac{m^2 b^2}{a^2} + \frac{P_y}{P_x} n^2 \right\}. \tag{7.48b}$$

Evidently the amplitude becomes infinite when $\Theta \to 0$. This occurs when

$$P_x = P_{x_{mn}} = \frac{\Phi_0 \pi^2 D_x}{b^2 \{(m^2 b^2/a^2 + (P_y/P_x)n^2\}}. \tag{7.56}$$

This is therefore the critical value of P_x which causes buckling in the mode m, n. It may be written also in the more familiar form:

$$P_{xmn} = \frac{\pi^2}{b^2} \frac{D_x}{g} K_3, \qquad (7.57\text{a})$$

where

$$K_3 = \frac{g\Phi_0}{(m^2 b^2/a^2) + n^2(P_y/P_x)}. \qquad (7.57\text{b})$$

For any particular sandwich and for any particular load ratio P_y/P_x the values of m and n must be chosen to provide the lowest possible value of P_{xmn}. There is no simple general rule which governs the choice of m and n except that when P_y is zero the smallest value of P_{xmn} is obtained by writing $n = 1$. That is, the panel buckles into one half-wave across the direction of loading.

Having fixed the value of n, K_3 may be plotted against a/b for the various values $m = 1, 2, 3, \ldots$; the envelope of the lowest of these curves provides the desired relationship between $(K_3)_{\min}$ and a/b.

Clearly it is not practicable to provide such plots in the general case because there are too many variables (such as D_x, D_{Qy}, v_x). Instead, a few cases of common interest will be considered.

Isotropic faces; isotropic antiplane core; critical value of P_x acting alone

Many panels with cores of expanded materials (e. g. expanded PVC) fall within this category, for which:

$$D_x = D_y = D_{xy}(1 + v_f)$$

and
$$\frac{B_{Qx}}{B_{Qy}} = \frac{D_{Qx}}{D_{Qy}} = 1.$$

Figure 7.4b shows a family of curves of K_3 plotted against the aspect ratio of the plate, a/b. Each curve is for a different value of D_x/B_{Qx} and the ripples in the curve correspond to different values of m inserted in the expression (7.57b). In every case P_y is zero, n is unity and the Poisson's ratio of the face material has been taken as 0·25.

The critical value of the edge load P_x is obtained by evaluating the stiffnesses D_x, D_{Qx} (Section 7.6), reading K_3 from Fig. 7.4b and inserting the result in equation (7.57a).

Isotropic faces; orthotropic antiplane core; critical value of P_x acting alone

Panels with honeycomb cores fall into this category, for which it has been assumed that:

$$D_x = D_y = D_{xy}(1+v_f)$$

and $$\frac{B_{Qx}}{B_{Qy}} = \frac{D_{Qx}}{D_{Qy}} = 0.4 \quad \text{or} \quad 2.5.$$

The ratios chosen for the two core shear stiffnesses are based on the results of tests on typical honeycomb cores, which show that the stiffness in one direction is often about two and a half times the stiffness in the other.

Figures 7.4a and 7.4c give values for K_3 in terms of b/a and the ratio D_x/B_{Qx} (not D_x/B_{Qy}).

Isotropic faces; corrugated core; critical value of P_x acting alone

For a corrugated core sandwich with the corrugations in the x-direction, the various plate stiffnesses are listed in Section 7.6 (iv). Figure 7.5 shows values of K_3 in terms of a/b and $D_y/b^2 D_{Qy}$ for four values of D_x/D_y ranging from 1·0 to 1·75. It is likely that most practical sandwiches will fall within these limits.

Figures 7.5 determine the critical value of the edge load P_x when it acts along the length of the corrugations; this is the most effective way of utilizing the panel but, for comparison, Fig. 7.6 shows values of K_3 when the load acts across the direction of the corrugations. Figure 7.6 is directly comparable with Fig. 7.5a.

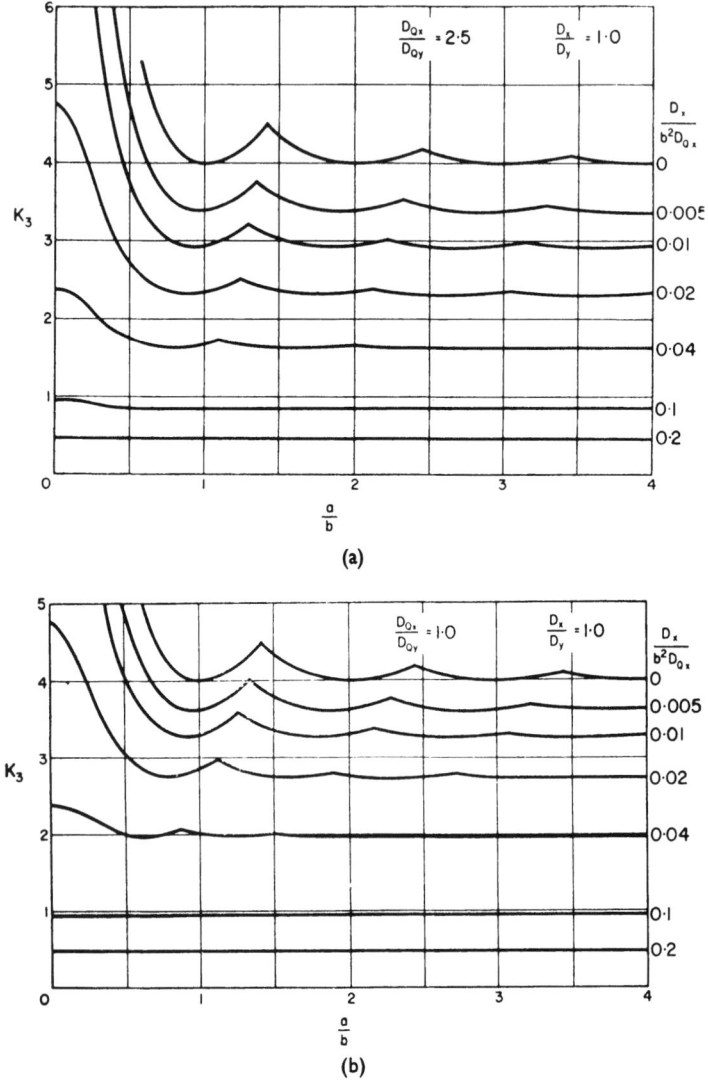

FIG. 7.4(a)–(c). Buckling coefficient K_3 in equation (7.57). Simply-supported sandwich panel with uniform edge load in x-direction. Isotropic thin faces and isotropic or orthotropic (*honeycomb*) core. $v_f = 0.25$.

(c)

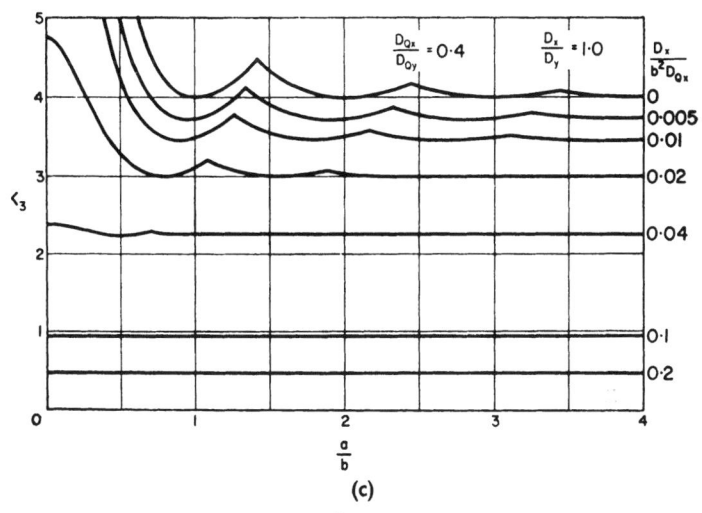

FIG. 7.4 (*cont.*).

7.9. Solution for Simply-supported Orthotropic Panel with Edge Loads and Uniform Transverse Load

It is easily shown that a transverse load of uniform intensity q may be represented by the series

$$q = \sum_{m=1}^{\infty} \sum_{n=1}^{\infty} q_{mn} \sin \frac{m\pi x}{a} \sin \frac{n\pi y}{b}, \quad (7.58a)$$

where
$$q_{mn} = \frac{16q}{mn\pi^2} \quad (m, n \text{ both odd}) \quad (7.58b)$$

$$= 0 \text{ (otherwise)}.$$

In order to obtain the effects due to the uniform pressure q, it is necessary to superimpose the separate effects of each term in the series (7.58a). These effects (displacements, bending and twisting moments, shear forces) have already been determined in Section 7.7. Furthermore, because only the odd modes are relevant, the

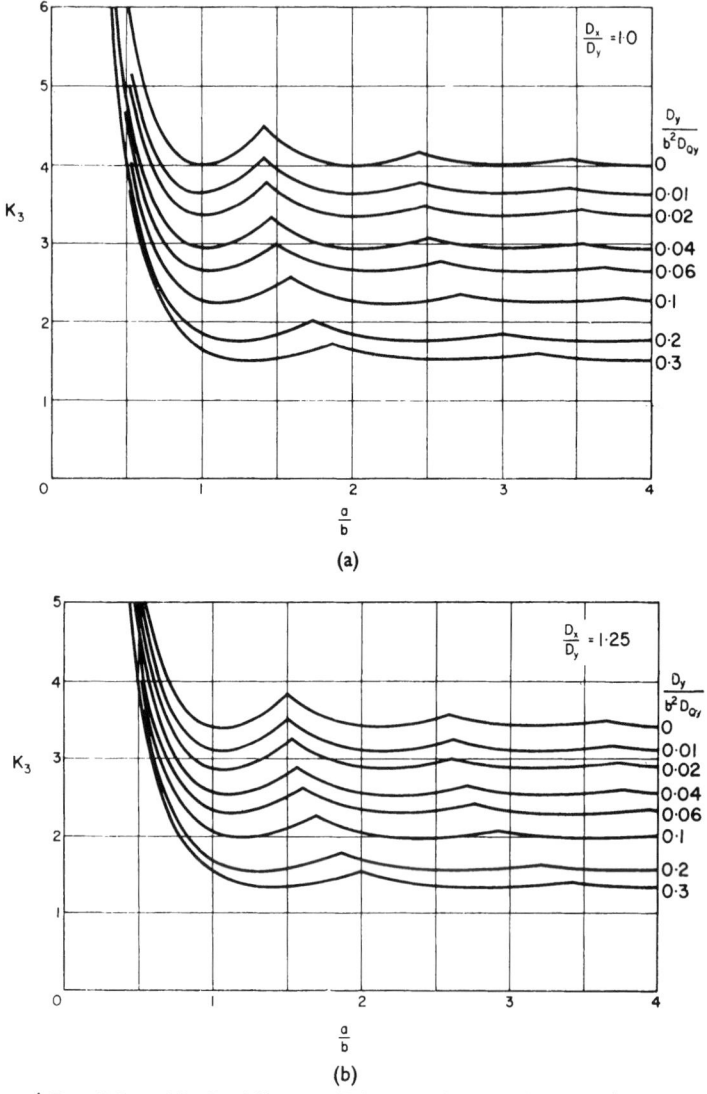

FIG. 7.5(a)–(d). Buckling coefficient K_3 in equation (7.57). Simply-supported sandwich panel with uniform edge load in x-direction. Isotropic thin faces and orthotropic (*corrugated*) core. Corrugations run in x-direction; $\nu_f = 0.25$; $D_{Qx} = \infty$.

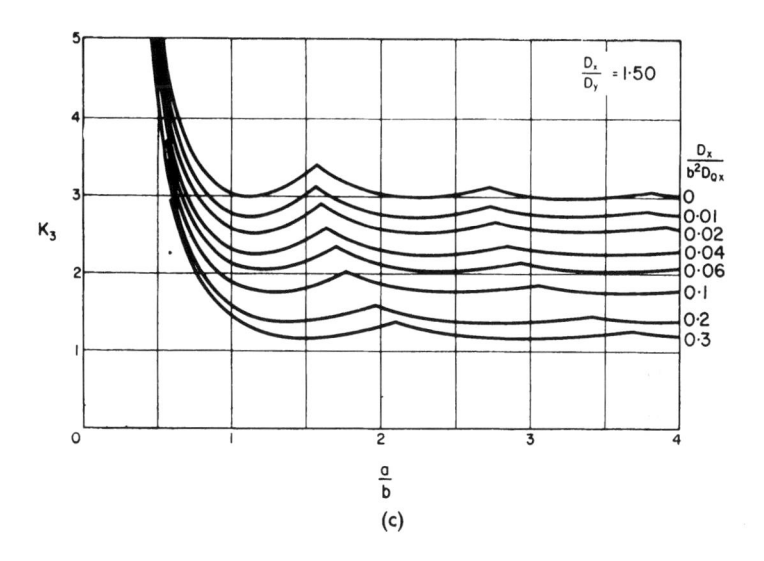

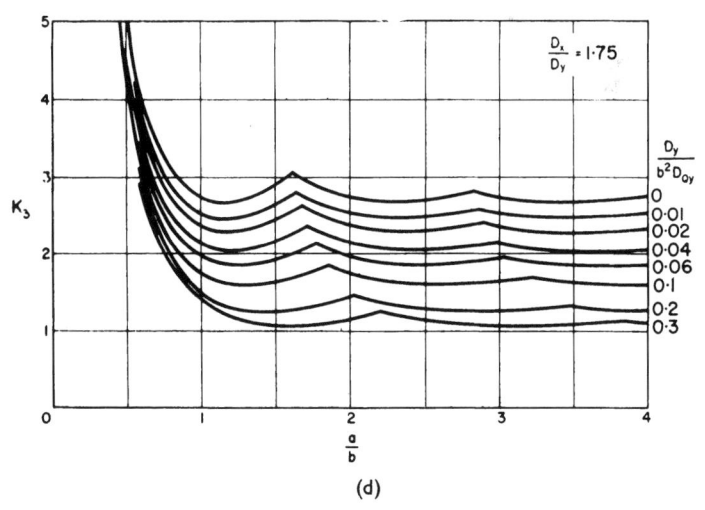

FIG. 7.5. (cont.).

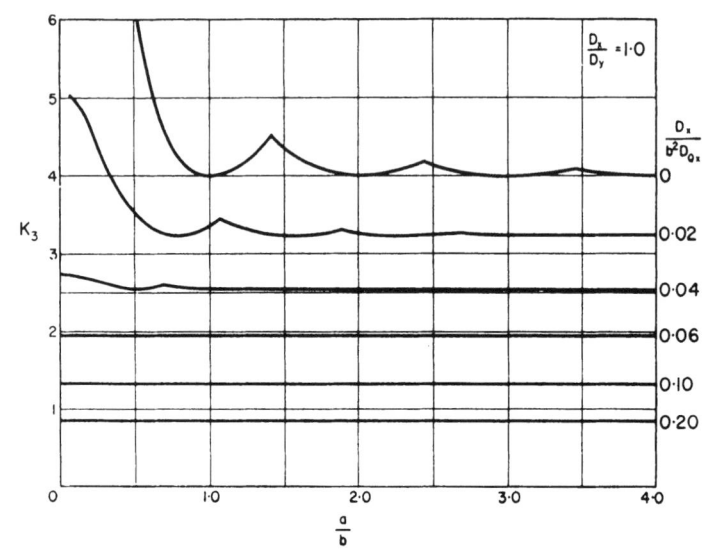

FIG. 7.6. Buckling coefficient K_3 in equation (7.57). Simply-supported sandwich panel with uniform edge load in x-direction. Isotropic thin faces and orthotropic *(corrugated)* core. Corrugations run in y-direction; $v_f = 0{\cdot}25$; $D_{Qy} = \infty$.

maximum effect in each mode occurs at the same position (x, y) in the plate. Thus the maximum effect due to the uniform pressure is obtained by superimposing the maximum effects in each odd mode.

From equations (7.49)–(7.55) (except (7.53)), therefore, the maximum effects due to a uniform transverse pressure q (in combination with compressive edge forces) are as follows:

$$w_{\max} = qb^4\beta_1/D_x, \tag{7.59}$$

$$(M_x)_{\max} = qb^2\beta_2, \tag{7.60}$$

$$(M_y)_{\max} = qb^2\beta_3, \tag{7.61}$$

$$(M_{xy})_{\max} = qb^2\beta_4, \tag{7.62}$$

$$(Q_x)_{\max} = qb\beta_5, \tag{7.63}$$

$$(Q_y)_{\max} = qb\beta_6, \tag{7.64}$$

where:

$$\beta_1 = \frac{16}{\pi^4} \sum \sum \frac{pq}{mn\Theta}, \tag{7.65a}$$

$$\beta_2 = \frac{16}{\pi^2 g} \sum \sum \left\{ \frac{\lambda' m^2 b^2}{a^2} + \mu' \nu_y n^2 \right\} \frac{pq}{mn\Theta}, \tag{7.65b}$$

$$\beta_3 = \frac{16}{\pi^2 g} \sum \sum \frac{D_y}{D_x} \left\{ \mu' n^2 + \lambda' \nu_x \frac{m^2 b^2}{a^2} \right\} \frac{pq}{mn\Theta}, \tag{7.65c}$$

$$\beta_4 = \frac{8}{\pi^2} \sum \sum \frac{D_{xy}}{D_x} \left(\frac{\lambda' + \mu'}{\Theta} \right) \frac{b}{a}, \tag{7.65d}$$

$$\beta_5 = \frac{16}{\pi^3} \sum \sum \frac{B_{Qx}(1-\lambda')}{D_x} \frac{b}{a} \frac{q}{n\Theta}, \tag{7.65e}$$

$$\beta_6 = \frac{16}{\pi^3} \sum \sum \frac{B_{Qy}(1-\mu')}{D_x} \frac{p}{m\Theta}. \tag{7.65f}$$

$p = (-1)^{(m-1)/2}$; $q = (-1)^{(n-1)/2}$; all summations are carried out for odd values of m and n.

The coefficients β_1–β_6 have been presented in graphical form for a square plate for several cases of interest as follows.

Isotropic faces; isotropic antiplane core

Many panels with cores of expanded materials (e.g. expanded PVC) fall within this category, for which:

$$D_x = D_y = D_{xy}(1 + \nu_f) \quad \text{and} \quad \frac{B_{Qx}}{B_{Qy}} = \frac{D_{Qx}}{D_{Qx}} = 1.$$

Figure 7.7 shows a family of curves for β_1–β_6 plotted against the ratio $D_x/b^2 D_{Qx}$. In every case the Poisson's ratio of the face material has been taken as 0·25.

The values of the panel stiffnesses may be obtained from Section 7.6.

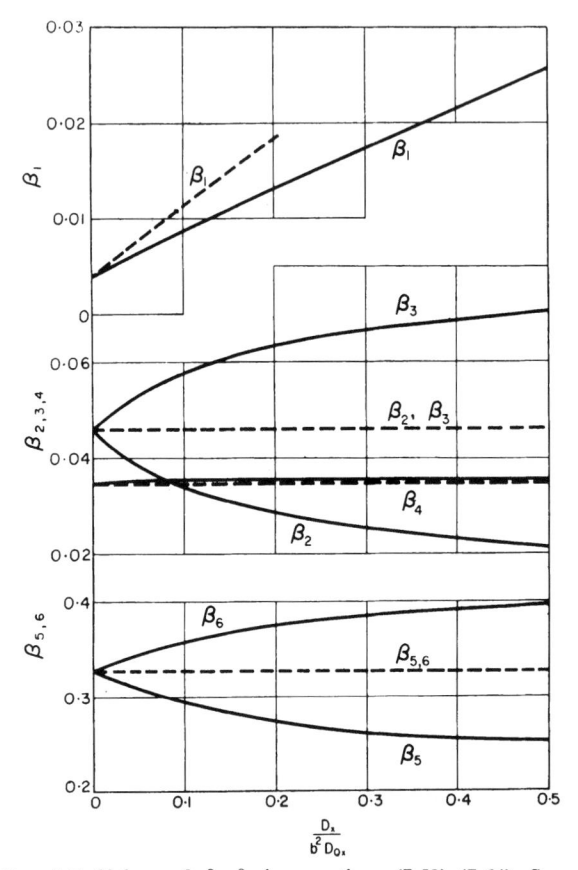

FIG. 7.7. Values of β_1–β_6 in equations (7.59)–(7.64). Square panel with isotropic thin faces. Broken lines are for isotropic core. Full lines are for orthotropic (*honeycomb*) core with $D_{Qx}/D_{Qy} = 0.4$. In every case $\nu_f = 0.25$ and the summations in equations (7.65) have been taken up to and including $m = n = 23$.

Isotropic faces; orthotropic antiplane core

Panels with honeycomb cores fall into this category, for which it has been assumed that:

$$D_x = D_y = D_{xy}(1+\nu_f) \quad \text{and} \quad \frac{B_{Qx}}{B_{Qy}} = \frac{D_{Qx}}{D_{Qy}} = 0.4.$$

The ratio of 0.4 is taken as typical for many honeycomb cores. Figure 7.7 shows a family of curves for β_1–β_6 plotted against the ratio $D_x/b^2 D_{Qx}$.

Isotropic faces; corrugated core

Figure 7.8 shows values of β_1–β_6 plotted against $D_y/b^2 D_{Qy}$. Curves are shown for values of D_x/D_y in the practical range, 1·0 to 1·75. The corrugations run in the x-direction.

Calculation of deflections and stresses

When the coefficients β_1–β_6 have been determined the deflections, bending moments and shear forces may be obtained at once from equations (7.59)–(7.64).

The stresses in a sandwich (other than a corrugated-core sandwich) may be obtained from the following formulae:

Face stresses (suffixes 1 and 2 refer to upper and lower faces):

$$\sigma_{x1} = M_x/dt_1; \quad \sigma_{x2} = M_x/dt_2.$$
$$\sigma_{y1} = M_y/dt_1; \quad \sigma_{y2} = M_y/dt_2.$$
$$\tau_{xy1} = M_{xy}/dt_1; \quad \tau_{xy2} = M_{xy}/dt_2.$$

Core stresses:

$$\tau_{zx} = Q_x/d; \quad \tau_{yz} = Q_y/d.$$

For corrugated-core panels the face stresses may be obtained approximately by the ordinary theory of bending, taking the appropriate second moments of area in each direction. However, there is no simple method of evaluating the shear stresses in the corrugated core; for this it is necessary to turn to the detailed analysis of Libove and Hubka.[12.4]

(a)

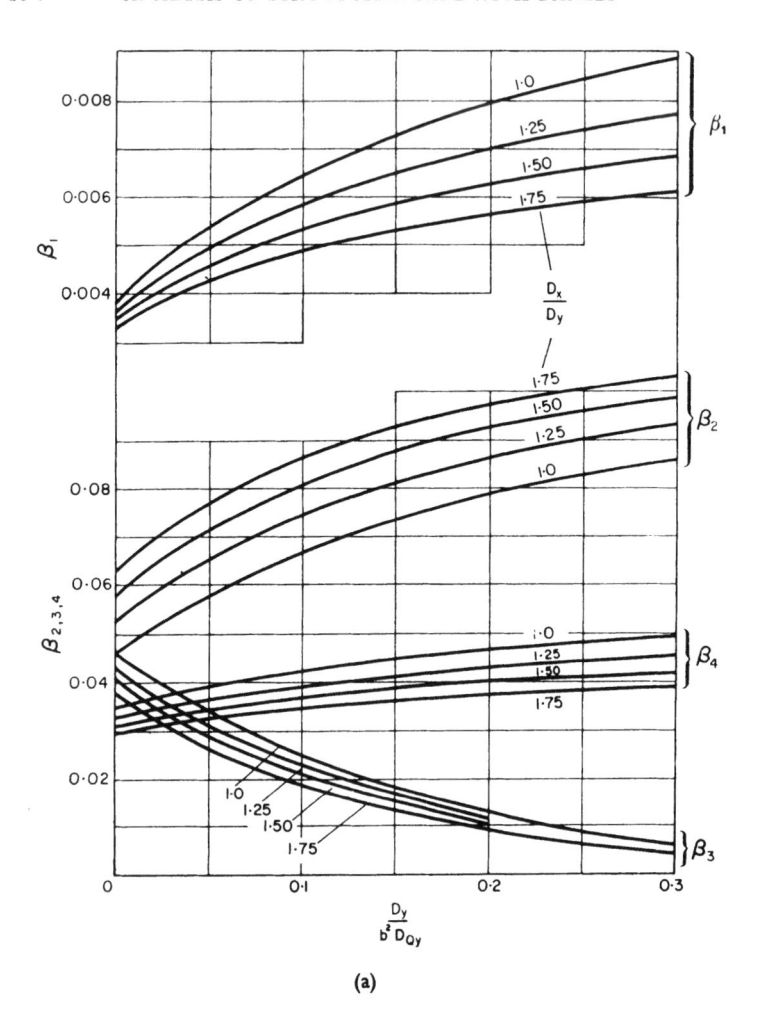

FIG. 7.8(a), (b). Values of β_1–β_6 in equations (7.59)–(7.64). Square panel with isotropic thin faces and orthotropic (*corrugated*) core. Corrugations run in x-direction; $\nu_f = 0.25$; $D_{Qx} = \infty$. Summations in equations (7.65) have been taken up to and including $m=n=23$. The ordinates for the β_1-curves for $D_x/D_y = 1.25, 1.5$ and 1.75 should be multiplied by 1.25, 1.5 and 1.75, respectively.

FIG. 7.8 (*cont.*).

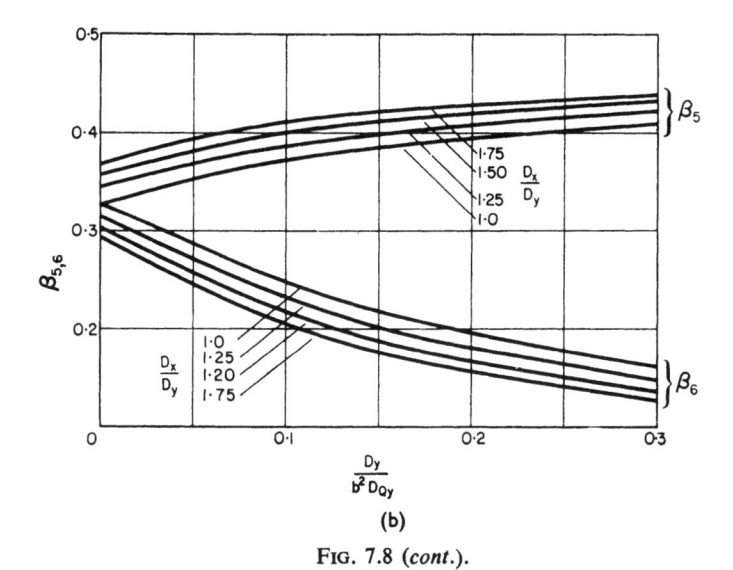

The values of the coefficients β in Figs. 7.7 and 7.8 have been obtained by taking the sums of the series in equations (7.65) up to and including the term $m = n = 23$. For a square plate which does not depart too far from perfect isotropy this is likely to give reasonably accurate results, with β_4 showing the poorest convergence.

CHAPTER 8

WRINKLING AND OTHER FORMS OF LOCAL INSTABILITY

8.1. Introduction

The compression faces of sandwich members are sometimes subject to a particular kind of instability aptly described as rippling or wrinkling, in which the wavelength of the buckled form is of the same order as the thickness of the core. Consider, for example, a pin-ended sandwich strut. Ordinary buckling theory (Chapters 3, 4) indicates that the lowest critical load is that which corresponds to a buckled form in which the half-wavelength is equal to the length of the strut. In suitable circumstances, however, short-wavelength wrinkling instability of the faces may occur at a still lower load. A special theory is needed to predict this behaviour, and the analysis which follows is based on the original work of Gough, Elam and De Bruyne,[10.1] with later work by Hoff and Mautner[9.1] and others.

The basic wrinkling theory is concerned only with local bending of the faces of the sandwich; the effect of membrane strains in the faces is completely neglected. For this reason it is entirely distinct from ordinary buckling theory, which is primarily concerned with membrane strains in the faces due to overall bending and buckling of the sandwich.

A link between the basic wrinkling theory and the ordinary buckling theory was forged by Williams, Leggett and Hopkins[8.1] with a unified strut analysis which reproduced the effects of wrinkling and overall instability and the interaction between them. The work of these authors (which is referred to later in the chapter) was later confirmed by Goodier[9.3, 10.2] in an exhaustive study of the same problem.

8.2. Long Strut Supported by a Continuous Elastic Isotropic Medium

It is convenient to begin with the very simple problem of an infinitely long strut attached to an elastic medium which extends to infinity on one side of the strut (Fig. 8.1). The strut and the medium are of width b in the y-direction; b is small, so that the system is in a state of plane stress ($\sigma_y = 0$). The strut is of rectangular section, of thickness t, and it represents the face of a sandwich member with a core of infinite thickness.

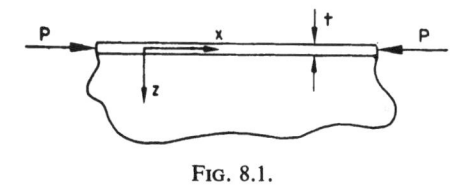

FIG. 8.1.

The differential equation of the strut is:

$$D \frac{d^4w}{dx^4} + P \frac{d^2w}{dx^2} = b\sigma_z, \qquad (8.1)$$

where D is the flexural rigidity of the strut, P is the axial thrust in the strut, w is the displacement in the z-direction and σ_z is the corresponding normal stress between the strut and the face of the supporting medium (σ_z is positive when tensile).

Suppose that the strut buckles into sinusoidal waves with a half-wavelength l, such that:

$$w = w_m \sin \frac{\pi x}{l}. \qquad (8.2)$$

The surface of the supporting medium must share this displacement (provided the face is thin) and it will be shown (Section 8.4) that the stress σ_z which is necessary to deform the medium in this manner may be written

$$\sigma_z = -\frac{a}{l} w_m \sin \frac{\pi x}{l}, \qquad (8.3)$$

where

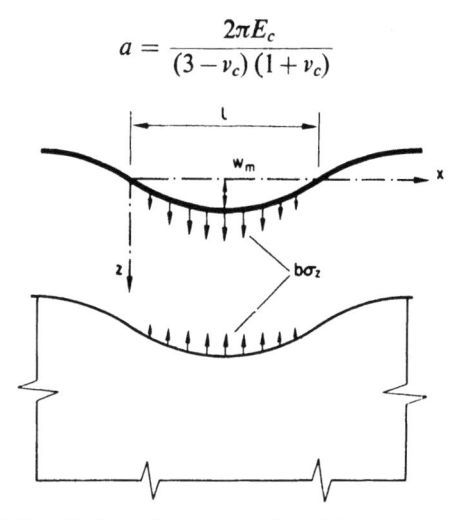

$$a = \frac{2\pi E_c}{(3 - v_c)(1 + v_c)}$$

FIG. 8.2. Tensile force $b\sigma_z$ per unit length between strut and supporting elastic medium.

and E_c, v_c are the modulus of elasticity and Poisson's ratio for the core. For a given half-wavelength l, the stress σ_z is proportional to the displacement w at any point x (Fig. 8.2).

Substitution for w and σ_z from equations (8.2) and (8.3) in (8.1) yields an expression in which ($w_m b \sin \pi x/l$) cancels, leaving the result:

$$D \frac{\pi^4}{l^4} - P \frac{\pi^2}{l^2} = -\frac{a}{l}. \qquad (8.4)$$

This equation defines the critical value of P which must exist to maintain the strut in a buckled condition. It is convenient to write

$$D = \frac{E_f t^3}{12}, \quad P = \sigma b t,$$

where E_f is the modulus of elasticity of the strut and σ is the *compressive* critical stress in the face. Rearranged, equation (8.4) appears as follows:

$$\sigma = \frac{\pi^2 E_f}{12}\left(\frac{t}{l}\right)^2 + \frac{a}{\pi^2}\left(\frac{l}{t}\right). \tag{8.5}$$

The critical stress is evidently a function of l/t, in the manner illustrated in Fig. 8.3. It is easy to show, by writing

$$\frac{d\sigma}{d(l/t)} = 0,$$

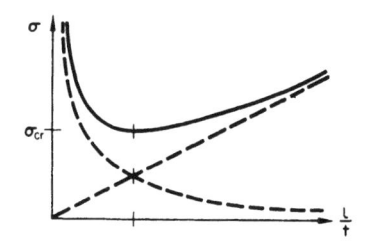

FIG. 8.3. Typical variation of critical stress with l/t (equation (8.5)). The broken lines illustrate the two terms on the right-hand side of equation (8.5).

that the minimum critical stress and the half-wavelength at which it occurs are given by the following expressions:

$$\sigma_{cr} = B_1 E_f^{\frac{1}{3}} E_c^{\frac{2}{3}} \quad \text{where} \quad B_1 = 3[12(3-\nu_c)^2(1+\nu_c)^2]^{-\frac{1}{3}}, \tag{8.6a}$$

$$\left(\frac{l}{t}\right)_{cr} = C\left(\frac{E_f}{E_c}\right)^{\frac{1}{3}} \quad \text{where} \quad C = \pi\left[(3-\nu_c)(1+\nu_c)/12\right]^{+\frac{1}{3}} \tag{8.6b}$$

Some values of the constants B_1, C are shown in Table 8.1 which shows that they are only slightly affected by variations in the Poisson's ratio of the supporting medium.

TABLE 8.1

v_c	B_1	C
0	0·630	1·98
0·25	0·575	2·07
0·30	0·570	2·09
0·5	0·545	2·13

The stresses in the core which are associated with the sinusoidal disturbance at the surface ($z = 0$) gradually decay as z increases; at a sufficiently large value of z they may be regarded as negligible. Consequently, the compression face of a sandwich beam or either face of a sandwich strut may be idealized in the manner of Fig. 8.1 provided the depth of the core is sufficiently great to allow the effects of a disturbance at one interface to die away before reaching the other interface. Equation (8.6a) may therefore be taken as a measure of the stress in the face at which wrinkling occurs, provided the core is thick.

If the core is not thick, there is some interaction between the faces on opposite sides of the core and equations (8.6) are modified to some extent. The modifications which arise in various circumstances are discussed briefly in the next section.

8.3. Wrinkling of Faces of Sandwiches with Isotropic Cores of Finite Thickness

The three principal cases of interest are illustrated in Fig. 8.4.

Case I represents a sandwich beam in which wrinkling is likely to occur only in the compression face; the tensile face is assumed to remain perfectly flat. Cases II and III represent respectively antisymmetrical and symmetrical wrinkling in a sandwich strut in which the two faces carry equal axial thrusts P.

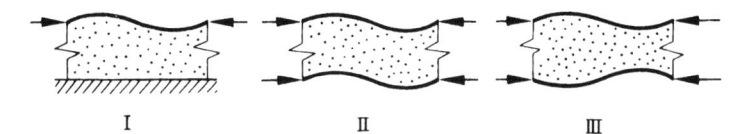

FIG. 8.4. Principal types of wrinkling instability.

In each of the three cases the transverse displacements of the upper face must satisfy the differential equation (8.1), and the displacements may be assumed to be sinusoidal in form (equation (8.2)). As in the previous section, the tensile normal stress σ_z at the interface is also sinusoidal in form, but the amplitude differs in each of the three cases. It is convenient to write:

$$\sigma_z = -\frac{w_m}{c} E_c \cdot \theta^2 f(\theta) \sin \frac{\pi x}{l}, \qquad (8.7)$$

where $\theta = \pi c/l$ and c is the thickness of the core. The function $f(\theta)$ is given in Table 8.2 for each case, together with the associated stress function which defines the state of stress in the core. The exact forms of these functions are not essential to an understanding of the present discussion. Their significance is discussed in Section 8.4, and further details may be found in papers by Gough, Elam and De Bruyne[10.1] (cases I and II) and by Hoff and Mautner[9.1] (case III).

Substitution of σ_z and w from equations (8.2) and (8.7) in (8.1) yields the following equation for the critical load P:

$$D\frac{\pi^4}{l^4} - P\frac{\pi^2}{l^2} = -\frac{bE_c}{c} \theta^2 f(\theta). \qquad (8.8)$$

A further rearrangement, with the substitutions $D = bE_f t^3/12$, $P = b\sigma t$, $\pi/l = \theta/c$, results in an expression for the critical *compressive* stress σ in the face.

$$\sigma = \frac{E_f}{12}\left(\frac{t}{c}\right)^2 \theta^2 + E_c\left(\frac{c}{t}\right) f(\theta). \qquad (8.9)$$

If there exists a value of θ (or l) at which the critical stress σ is a minimum, then that value of θ must satisfy the equation:

$$\frac{d\sigma}{d\theta} = \frac{E_f}{6}\left(\frac{t}{c}\right)^2 \theta + E_c\left(\frac{c}{t}\right)f'(\theta) = 0. \quad (8.10)$$

This cannot be solved directly for θ; instead it may be written in the equivalent form:

$$\frac{t}{c}\left\{\frac{E_f}{E_c}\right\}^{1/3} = -\left\{\frac{6f'(\theta)}{\theta}\right\}^{1/3} = \varrho, \text{ say.} \quad (8.11)$$

Equation (8.11) may be used to plot ϱ against θ for all circumstances in which a true minimum value of σ exists. For a given sandwich ϱ can be evaluated and the corresponding value of θ can be read from the graph and inserted in equation (8.9) to provide the minimum critical wrinkling stress. It will be found convenient to write equation (8.9) in an alternative form:

$$\sigma = B_1 E_f^{1/3} E_c^{2/3}, \quad (8.12a)$$

where

$$B_1 = \varrho^2\theta^2/12 + f(\theta)/\varrho. \quad (8.12b)$$

The value of the constant B_1 in each of the three cases is discussed in detail below.

CASE I. RIGID BASE

A graph of l/c against ϱ, based on equation (8.11), is shown in Fig. 8.5. The ratio l/c has been used instead of θ as a matter of convenience ($l/c = \pi/\theta$), and the appropriate value of $f'(\theta)$ has been obtained from Table 8.2. It may be seen that whatever the value of ϱ there always exists a buckled half-wavelength l at which the critical stress σ is minimum. The value of the constant B_1 which defines this minimum critical stress is also shown in Fig. 8.5.

Many practical sandwiches with rather thin faces fall in the range $\varrho < 0.25$; in that case the value of B_1 may be taken as 0·575

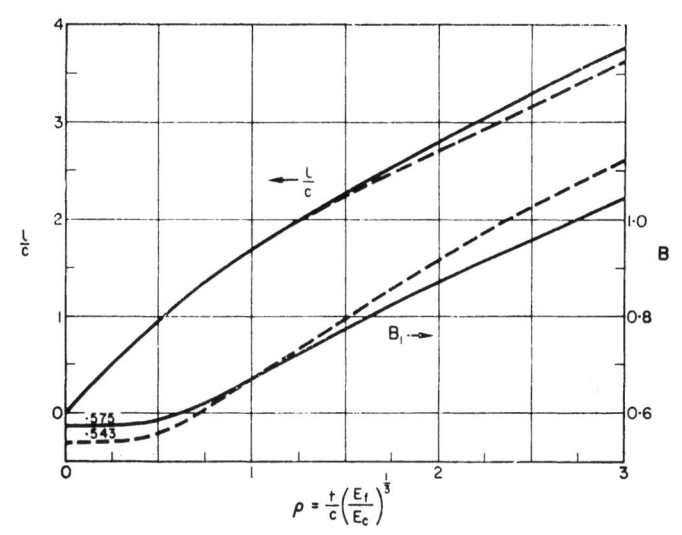

FIG. 8.5. Wrinkling (case I). l is the half-wavelength of the buckled face; B_1 is the buckling coefficient in equation (8.12a). Full and broken lines are for $\nu_c = 0.25$ and 0.5, respectively. (Based on Fig. 5 of ref. 10.1. By permission of the Royal Aeronautical Society.)

or 0.543 for $\nu_c = 0.25$ or 0.5 respectively. When ϱ is larger than 0.25 these values underestimate the wrinkling stress by a considerable margin.

CASE II. ANTISYMMETRICAL WRINKLING

A graph of l/c against ϱ, based on equation (8.11), is shown in Fig. 8.6; again, the appropriate value of $f'(\theta)$ has been derived from Table 8.2. The curves exist only for values of ϱ which are less than a limiting value

$$\left\{\frac{1-\nu_c}{8(1+\nu_c)}\right\}^{\frac{1}{3}}.†$$

† The index is incorrectly given as $1/2$ in ref. 10.1.

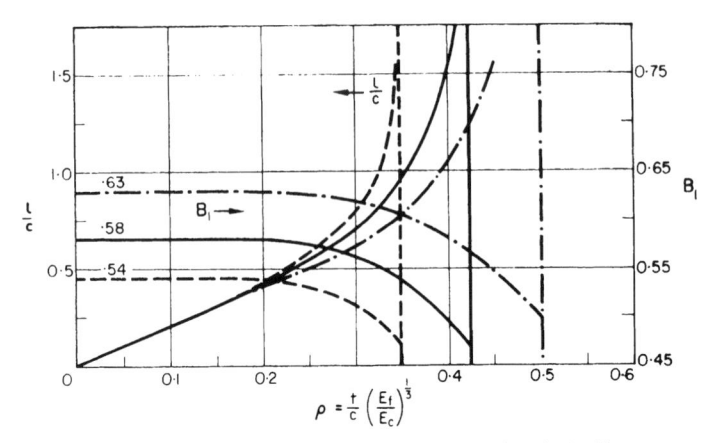

FIG. 8.6. Antisymmetric wrinkling (case II). l is the half-wavelength of the buckled face; B_1 is the buckling coefficient in equation (8.12a). Chain, full and broken lines are for $v_c = 0$, 0·25, 0·5, respectively. (Based on Figs. 13 and 14 of ref. 10.1. By permission of the Royal Aeronautical Society.)

When ϱ lies below this value, therefore, there exists a buckled half-wavelength l at which the critical stress σ is minimum. The value of the constant B_1 which defines this minimum critical stress is also shown in Fig. 8.6.

For $\varrho < 0.20$, the value of B_1 may be taken as 0·63, 0·58 or 0·54 for $v_c = 0$, 0·25 or 0·5 respectively, but these values are slightly non-conservative for larger values of ϱ.

When ϱ is greater than the limiting value stated above, there is no solution to equation (8.10) and there is no finite value of θ (or l) at which the critical stress is minimum. Instead the critical stress decreases continuously as the half-wavelength is increased. In these circumstances failure may be caused by overall instability of the Euler type, modified by shear deformations in the core (Chapters 3 and 4). In other words, antisymmetrical wrinkling does not occur when ϱ exceeds the limiting value.

TABLE 8.2

Case I
$$f(\theta) = \frac{2}{\theta} \frac{(3-\nu_e)\sinh\theta\cosh\theta + (1+\nu_e)\theta}{(1+\nu_e)(3-\nu_e)^2 \sinh^2\theta - (1+\nu_e)^3 \theta^2}.$$

Case II
$$f(\theta) = \frac{2}{\theta} \frac{\cosh\theta - 1}{(1+\nu_e)(3-\nu_e)\sinh\theta + (1+\nu_e)^2 \theta}.$$

Case III
$$f(\theta) = \frac{2}{\theta} \frac{\cosh\theta + 1}{3\sinh\theta - \theta} \qquad (\nu_e = 0).$$

Case I (face at $z = 0$; rigid base at $z = +c$)
$$\phi = -(\sigma_z)_{\max}\left\{\frac{l^2}{\pi^2}\sin\frac{\pi x}{l}\right\}\left\{\cosh\psi - \frac{1+\nu_e}{2}\cdot\psi\sinh\psi\right.$$
$$\left. - \frac{(1+\nu_e)(3-\nu_e)\psi\cosh\psi\sinh^2\theta - \sinh\psi[2(3-\nu_e)\sinh^2\theta - (1+\nu_e)^2\theta^2]}{2(3-\nu_e)\sinh\theta\cosh\theta + (1+\nu_e)\theta}\right\}.$$

Case II (faces at $z = \pm c/2$)
$$\phi = -(\sigma_z)_{\max}\left\{\frac{l^2}{2\pi^2}\frac{\sin\pi x/l}{\sinh^2\theta/2}\right\}\left\{\sinh\psi\left[2\sinh\frac{\theta}{2} + (1+\nu_e)\frac{\theta}{2}\cosh\frac{\theta}{2}\right]\right.$$
$$\left. - (1+\nu_e)\psi\cosh\psi\sinh\frac{\theta}{2}\right\}.$$

Case III (faces at $z = \pm c/2$)
$$\phi = +(\sigma_z)_{\max}\left\{\frac{l^2}{2\pi^2}\frac{\sin\pi x/l}{\cosh^2\theta/2}\right\}\left\{\psi\sinh\psi\cosh\frac{\theta}{2}\right.$$
$$\left. - \left[\frac{\theta}{2}\sinh\frac{\theta}{2} + 2\cosh\frac{\theta}{2}\right]\cosh\psi\right\},$$

$\theta = \dfrac{\pi c}{l}$; $\psi = \dfrac{\pi z}{l}$; σ_z = normal stress at interface (tensile).

CASE III. SYMMETRICAL WRINKLING

A graph of l/c against ϱ, based on equation (8.11) with the appropriate value of $f'(\theta)$ derived from Table 8.2, is shown in Fig. 8.7. Whatever the value of ϱ there always exists a buckled half-wavelength l at which the critical stress σ is minimum. The value of B_1 which defines this minimum critical stress is also shown in Fig. 8.7.

For $\varrho < 0.25$, and for zero Poisson's ratio, the value of B_1 is approximately 0·63. It will be observed that for larger values of ϱ, B_1 is always larger than 0·63. In the antisymmetric case, how-

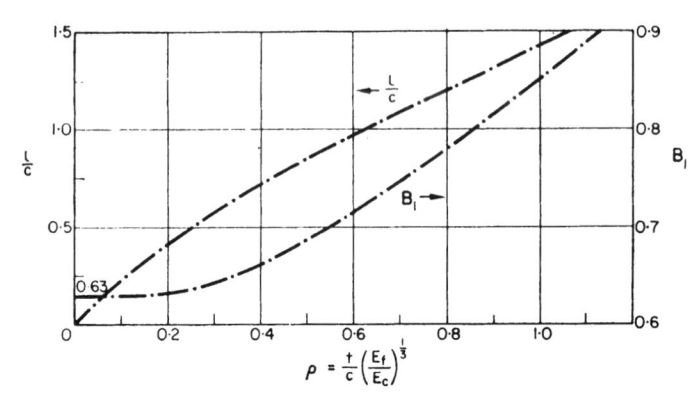

FIG. 8.7. Symmetric wrinkling (case III). l is the half-wavelength of the buckled face; B_1 is the buckling coefficient in equation (8.12a). Both curves are for $v_c = 0$.

ever, when wrinkling is possible it occurs at values of B_1 which are never larger than 0·63 ($v_c = 0$). It is therefore possible to draw the following conclusions:

(i) When ϱ is very small (<0·2, say) symmetrical and unsymmetrical wrinkling are equally likely to occur.
($B_1 = 0.63$ for $v_c = 0$.)

(ii) When $0 \cdot 2 < \varrho < \left\{ \dfrac{1-v_c}{8(1+v_c)} \right\}^{\frac{1}{3}}$, antisymmetric wrinkling occurs at a lower stress than does symmetric wrinkling. The value of B_1 may be read from Fig. 8.6.

(iii) When $\varrho > \left\{ \dfrac{1-v_c}{8(1+v_c)} \right\}^{\frac{1}{3}}$, antisymmetric wrinkling cannot occur but symmetric wrinkling is possible at values of B_1 given by Fig. 8.7.

(iv) The value of B_1 for antisymmetric wrinkling at $\varrho = \left\{ \dfrac{1-v_c}{8(1+v_c)} \right\}^{\frac{1}{3}}$ (Fig. 8.6) is always conservative for all types of wrinkling instability. This minimum value is 0·500, 0·47, or 0·47 for Poisson's ratios of 0, 0·25 and 0·5 respectively.

(v) If the strut is sufficiently long, overall instability of the type discussed in Chapters 3 and 4 will occur at a stress lower than the wrinkling stress.

8.4. Behaviour of the Core during Wrinkling

In Section 8.2 a long continuous strut was supported and restrained on one side by an elastic isotropic medium which extended to infinity in the positive direction of z. It was stated that a sinusoidal z-displacement of the surface of the medium (equation (8.2), Fig. 8.2) would be associated with a sinusoidal normal stress σ_z at the surface (equation (8.3)). It is now necessary to verify that this is so.

Stresses in the elastic isotropic medium may be defined by a stress function, $\phi(x, z)$, which must satisfy the following equation:

$$\left\{\frac{\partial^4}{\partial x^4} + 2\frac{\partial^4}{\partial x^2 \partial z^2} + \frac{\partial^4}{\partial z^4}\right\}\phi = 0. \tag{8.13}$$

The stresses and strains at (x, z) may be determined from the stress function ϕ as follows:

$$\sigma_x = \frac{\partial^2 \phi}{\partial z^2}; \quad \sigma_z = \frac{\partial^2 \phi}{\partial x^2}; \quad \tau_{zx} = -\frac{\partial^2 \phi}{\partial x \partial z}, \tag{8.14}$$

$$e_x = \frac{\partial u}{\partial x} = \frac{1}{E_c}(\sigma_x - \nu_c \sigma_z); \quad e_z = \frac{\partial w}{\partial z} = \frac{1}{E_c}(\sigma_z - \nu_c \sigma_x). \tag{8.15}$$

The displacements u and w may be obtained by integration of equation (8.15).

The stress function defined by equation (8.16) may be used in equations (8.14) and (8.15) to verify the results listed in (8.17).

$$\phi = A \sin \frac{\pi x}{l} (1 - Bz) \exp\left(-\frac{\pi z}{l}\right), \qquad (8.16)$$

$$z = 0 \begin{cases} \sigma_z = -A \dfrac{\pi^2}{l^2} \sin \dfrac{\pi x}{l}, & (8.17a) \\[6pt] \tau_{zx} = A \dfrac{\pi}{l} \cos \dfrac{\pi x}{l} \left(B + \dfrac{\pi}{l}\right), & (8.17b) \\[6pt] e_x = \dfrac{A}{E_c} \dfrac{\pi}{l} \sin \dfrac{\pi x}{l} \left[2B + \dfrac{\pi}{l}(1 + v_c)\right], & (8.17c) \\[6pt] w = \dfrac{A}{E_c} \sin \dfrac{\pi x}{l} \left[\left(\dfrac{\pi}{l} - B\right) + v_c\left(\dfrac{\pi}{l} + B\right)\right]. & (8.17d) \end{cases}$$

In Section 8.2 the strut which was attached to the surface of the medium was permitted to deflect in the z-direction only; there were to be no x-displacements. Consequently the x-displacements at the surface of the medium are zero, and so are the strains e_x at the surface (equation (8.17c)). If e_x is to vanish, the constant B must take the value:

$$B = -\frac{\pi}{2l}(1 + v_c). \qquad (8.18)$$

If the amplitude of the z-displacement is denoted by w_m and the expression (8.18) for B is substituted in equations (8.17), the following results are obtained:

$$\sigma_z = -\frac{w_m}{l} \frac{2\pi E_c}{(3 - v_c)(1 + v_c)} \cdot \sin \frac{\pi x}{l}, \qquad (8.19a)$$

$$\tau_{zx} = \frac{w_m}{l} \frac{\pi E_c (1 - v_c)}{(3 - v_c)(1 + v_c)} \cos \frac{\pi x}{l}, \qquad (8.19b)$$

$$e_x = 0, \qquad (8.19c)$$

$$w = w_m \sin \frac{\pi x}{l}. \qquad (8.19d)$$

Equations (8.19a) and (8.19d) are identical with equations (8.3) and (8.2). Equation (8.19c) verifies that there are no displacements at the surface in the x-direction, but equation (8.19b) shows that there exists a sinusoidal shear stress at the interface. The effect of this shear stress is to introduce small periodic fluctuations in the axial thrust in the strut; these small fluctuations have been neglected in the formation of the differential equation (8.1). The shear stress at the interface also introduces bending moments into the strut because of its eccentricity with respect to the centre-line of the strut; this effect has been ignored by supposing that the strut is very thin. Some error may be introduced when the face is thick.

The stress functions ϕ listed in Table 8.2 may be used in a completely analogous manner to describe the behaviour of the isotropic cores in cases I, II and III, Section 8.3, and to define the function $f(\pi c/l)$ in equation (8.7).

8.5. The Winkler Hypothesis

In the analysis of the behaviour of a long strut or beam supported by a continuous elastic medium it is sometimes supposed that the medium can be replaced by a set of closely-spaced springs, as in Fig. 8.8, without serious error. This is sometimes described as the Winkler hypothesis.

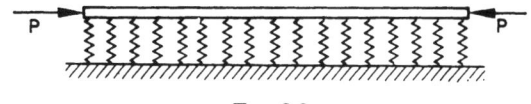

FIG. 8.8.

Suppose that the stiffness of the springs is denoted by a constant λ, which represents the force needed to displace the springs in a unit area of the xy-plane through a unit distance in the z-direction. The interface may be given the displacement:

$$w = w_m \sin \frac{\pi x}{l}. \qquad (8.2)$$

The corresponding normal stress at the interface (tension positive) is therefore

$$\sigma_z = -\lambda w_m \sin \frac{\pi x}{l}. \tag{8.20}$$

If, now, the system in Fig. 8.8 is assumed to be equivalent to that in Fig. 8.4(I), it is necessary for equation (8.20) to approximate to equation (8.7) (with the appropriate value for $f(\theta)$). Thus:

$$\lambda = \frac{E_c}{c} \theta^2 f(\theta). \tag{8.21}$$

The right-hand side is a function of θ, or $\pi c/l$, and it is not possible to select one value of λ which is appropriate for displace-

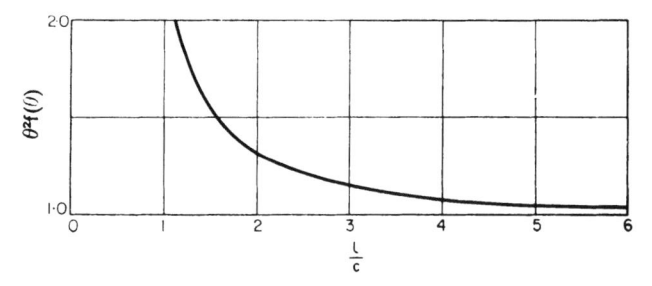

FIG. 8.9. Value of $\theta^2 f(\theta)$ in equation (8.21).

ments of any wavelength. However, a graph of $\theta^2 f(\theta)$ against l/c is shown in Fig. 8.9 for $v_c = 0$. It is clear that, provided $l/c > 3$, $\theta^2 f(\theta)$ is very close to unity.

Any disturbance of the beam or strut in the z-direction may be broken down into Fourier sine components of different wavelengths. For each component with a half-wavelength greater than $3c$, the supporting medium provides restraint which is indistinguishable from closely-spaced springs of stiffness;

$$\lambda = \frac{E_c}{c}. \tag{8.22}$$

Consequently, by the principle of superposition, provided the displacement components with half-wavelengths of less than $3c$ are insignificant, equation (8.22) can be used to describe the overall behaviour of the elastic supporting medium.

Equation (8.22) is the result which would have been obtained from elementary considerations, that is to say, by dividing the core into narrow columns in the z-direction, each of height c and of unit cross-sectional area. Each column represents a spring of stiffness $\lambda = E_c/c$. It is evident that a model of this kind takes into account the stiffness of the core in direct compression in the z-direction, but completely neglects the shear stiffness in the zx-plane. It is for this reason that the Winkler hypothesis becomes inadequate for deformations of short wavelength, in which shear deformations of the core become important.

The Winkler hypothesis cannot be applied to the infinite medium in Figs. 8.1 and 8.2. In this case it is necessary to compare equations (8.3) and (8.20). For $v_c = 0$ this implies that:

$$\lambda = \frac{2\pi E_c}{3l}. \tag{8.23}$$

Obviously there is no range of values of l for which λ may usefully be assumed constant.

8.6. Stability of Faces Attached to Antiplane Core with Infinite Stiffness Perpendicular to the Faces

In very many analyses of sandwich behaviour the assumptions are made that the core is antiplane† and that it possesses infinite stiffness in the z-direction, perpendicular to the faces. It is therefore of some interest to examine the implications of such assumptions for the wrinkling of the faces.

---

† i.e. the core carries no stresses in the xy-plane; the shear stresses in the zx- and yz-planes are constant through the depth of the core.

Because of the infinite transverse core stiffness, the distance between the faces remains constant at all times and the wrinkling modes I and III (Fig. 8.4) are not possible. The possibility of antisymmetrical wrinkling (type II) remains to be investigated. Figure 8.10 illustrates the forces acting on an element of the core of dimensions $dx \times b \times c$ during antisymmetrical wrinkling. The out-of-balance shear force $(d\tau/dx)bcdx$ is equilibrated by a tensile

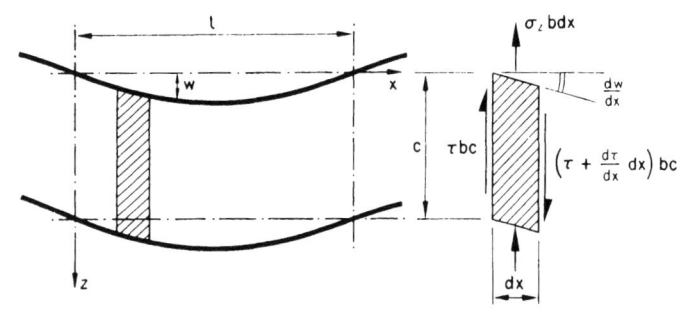

FIG. 8.10. Forces acting on an element of antiplane core.

force $\sigma_z bdx$ applied at the upper interface and an equal compressive force applied at the lower interface. Consequently,

$$\sigma_z = \frac{c}{2} \frac{d\tau}{dx}. \tag{8.24}$$

The shear stress τ may be expressed in terms of the shear strain, which in turn is equal to dw/dx. If the z-displacement is sinusoidal, as in equation (8.2), then:

$$\sigma_z = -\frac{cG_c}{2} w_m \frac{\pi^2}{l^2} \sin \frac{\pi x}{l}. \tag{8.25}$$

Substitution for σ_z and w in the differential equation (8.1) for the face yields the following result for the critical stress in the face, σ:

$$\sigma = \frac{\pi^2 E_f t^2}{12 l^2} + \frac{G_c}{2} \frac{c}{t}. \tag{8.26a}$$

The total thrust supported by the two faces of the sandwich is:

$$2bt\sigma = 2\left\{\frac{\pi^2 E_f t^3 b}{12l^2}\right\} + G_c bc. \tag{8.26b}$$

Evidently the critical load is least when l is as large as possible. In a practical pin-ended strut the largest possible value for l is the actual length of the strut. On the right-hand side of equation (8.26b), therefore, the first term represents the sum of the Euler loads of the faces, buckling as independent struts; the second term represents the shear buckling load of the core.

Equation (8.26b) represents a special case of the strut equation (3.13) in which the Euler load P_E of the strut as a whole is infinite. This coincidence is to be expected because, in the analysis of the present chapter, it has been assumed so far that the membrane strains in the faces are zero; overall Euler-type instability has therefore been excluded from consideration.

For cores of the type under discussion there is no particular half-wavelength which corresponds to a true minimum value of the critical stress σ; therefore wrinkling does not occur.

8.7. Initial Irregularities of the Faces

In the preceding sections it has been assumed that the faces are perfectly straight and axially loaded. In such circumstances the faces remain straight under load until the stress in them reaches some critical value. When this happens, the faces are in a state of neutral equilibrium and a buckled configuration becomes possible in which the faces and the core suffer elastic deformations. This classical type of wrinkling failure does not depend in any way on a failure or rupture of the core or the faces.

A different type of failure may occur if the faces are not perfectly straight and axially loaded. As soon as load is applied, its eccentricity induces bending moments in the faces which therefore bend locally and their irregularity increases. The eccentricity of the end load is thereby increased in a vicious circle. As a consequence, the initial irregularities of the faces are magnified considerably by the action of the end load. The core is strained by these deformations and normal stresses (tensile and compressive) develop at the interfaces between the faces and the core. Eventually a point is reached at which either the adhesive or the core itself fails in tension.

Theoretically it is also possible for the core to fail in compression or for the face to fail in bending, but the tensile failure at the interface is certainly much more common.

The effect may be studied by assuming that when unloaded the face has a z-displacement w_0, which increases to a total value w under load, and that both displacements are sinusoidal:

$$w_0 = w_{om} \sin \frac{\pi x}{l}, \qquad (8.27)$$

$$w = w_m \sin \frac{\pi x}{l}. \qquad (8.2)$$

The tensile stress (σ_z) which develops at the interface between face and core is given by a modified form of equation (8.7) on the assumption that it vanishes when the system is unloaded; that is, when w_m is equal to w_{om}:

$$\sigma_z = -(w_m - w_{om}) \frac{E_c}{c} \theta^2 f(\theta) \sin \frac{\pi x}{l}; \qquad \theta = \frac{\pi c}{l}. \qquad (8.28)$$

The differential equation for the face is a modified form of equation (8.1). The modification arises because the bending deformation of the face is the difference between the final displacement w and the initial value w_0. The second term of the equation is unchanged because the eccentricity of the load end P is w, not $(w - w_0)$.

$$\frac{E_f b t^3}{12} \frac{d^4}{dx^4}(w - w_0) + b\sigma t \frac{d^2 w}{dx^2} = b\sigma_z. \qquad (8.29)$$

Substitution for w_0, w and σ_z from equations (8.27), (8.2) and

WRINKLING AND OTHER FORMS OF LOCAL INSTABILITY 175

(8.28) provides an equation which can be rearranged to express the final displacement amplitude in terms of the initial amplitude:

$$w_n = \frac{w_{om}}{1-\sigma/\sigma_{cr}}. \tag{8.30}$$

In this equation σ again represents the mean stress in the face; σ_{cr} is the critical value of this stress at which pure wrinkling of a perfectly straight axially-loaded face would occur; σ_{cr} is defined by equation (8.12) for any particular half-wavelength l.

Substitution for w_m from equation (8.30) in equation (8.28) determines the tensile interface stress σ_z, the amplitude of which is

$$\sigma_{zm} = -\frac{E_c}{c}\theta^2 f(\theta)\frac{\sigma}{\sigma_{cr}}\frac{w_{om}}{1-\sigma/\sigma_{cr}}. \tag{8.31}$$

If the sign is neglected, this is also the value of the maximum tensile interface stress associated with an initial irregularity of amplitude w_{om} and half-wavelength l. At failure it may be assumed that this maximum stress is equal to the ultimate tensile strength (σ_u, say) of the adhesive (or the core) at the interface, in the z-direction. The face stress σ at which failure occurs may be obtained by solving equation (8.31) for σ:

$$\sigma = \left\{\frac{E_f}{12}\left(\frac{t}{c}\right)^2\theta^2 + E_c\left(\frac{c}{t}\right)f(\theta)\right\}\left\{1+\frac{E_c}{c}\frac{w_{om}}{\sigma_u}\theta^2 f(\theta)\right\}^{-1}. \tag{8.32}$$

This equation cannot be put to practical use to determine the stress at which failure occurs unless something is known about w_{om} and σ_u. Williams[28.1] has suggested that w_{om} is proportional to the wavelength and Wan[9.2] that it is proportional to the square of the wavelength and the inverse of the face thickness. In a thorough investigation of this problem, Norris et al.[6.3] assumed that for a given face thickness w_{om} is proportional to the core thickness and inversely proportional to the modulus of elasticity of the core in the z-direction (E_c in this case). This assumption was intended to account for irregularities caused during fabrication, in which substantial pressures may be applied to the faces of the sandwich. The flattening of soft spots in the core, and the consequent denting of the faces, would be greater for thick cores of low modulus. If it is also supposed that σ_u is constant for any particular form of construction, it is possible to write:

$$w_{om} = \frac{kc\sigma_u}{E_c}, \qquad (8.33)$$

where k is a non-dimensional constant.

Equation (8.32) now appears as follows:

$$\sigma = B_2 E_f^{1/3} E_c^{2/3}, \qquad (8.34a)$$

where

$$B_2 = \left\{ \frac{\varrho^2 \theta^2}{12} + \frac{f(\theta)}{\varrho} \right\} \{1 + k\theta^2 f(\theta)\}^{-1} \qquad (8.34b)$$

and

$$\varrho = \frac{t}{c} \left(\frac{E_f}{E_c} \right)^{\frac{1}{3}}. \qquad (8.11)$$

Equations (8.34a) and (8.12a) are similar in form and they are identical when $k = 0$, signifying the absence of initial irregularities. The face stress σ at which tensile failure occurs at the interface is now defined by equation (8.34) as a function of θ, or of the half-wavelength of the initial irregularity. The worst possible situation occurs when B_2 is minimized with respect to θ; that is, when the initial irregularity happens to have the least favourable wavelength.

The equation $dB_2/d\theta = 0$ can be solved for pairs of values of ϱ and θ which can be inserted in equation (8.34) to provide the smallest value of B_2. Figure 8.11 shows values of B_2 plotted against $1/\varrho$ for various values of k; the Poisson's ratio v_c for the core has been taken as 0·25, but minor changes in the value of v_c are not likely to affect the curves to a great extent. The curve $k = 0$ corresponds exactly to the curve for B_1 ($v_c = 0·25$) in Fig. 8.6.

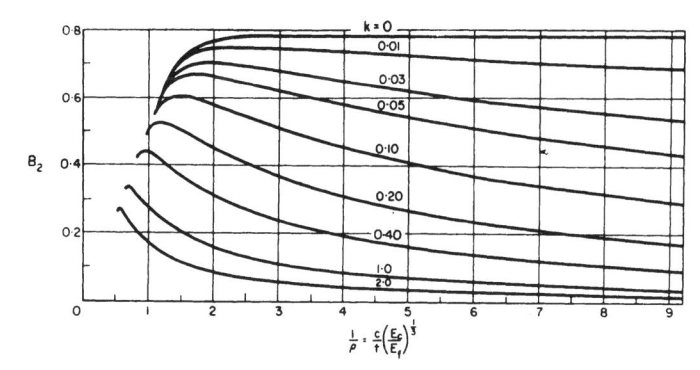

FIG. 8.11. Value of B_2 in equation (8.34). Isotropic core, $v_e = 0.25$. (After fig. 16 of ref. 6.3. By permission of the U.S. Forest Products Laboratory.)

It is important to understand the limitations of Fig. 8.11. It is based on the following assumptions:

(i) The ultimate tensile strength of the interface is constant.
(ii) The amplitude of initial irregularities is
   (a) independent of the wavelength of the irregularities,
   (b) inversely proportional to the core modulus E_c,
   (c) proportional to the core thickness c.
(iii) Different values of k may be appropriate for different face thicknesses.

If these conditions are fulfilled, then compression tests on a series of short sandwich columns of similar construction, with similar faces, but with different core thicknesses and core moduli should produce a plot of failure stress σ against ϱ which corresponds to a particular value of k in Fig. 8.11. If condition (iib) is not fulfilled it might be found that different values of k are appropriate for different core moduli. If condition (iia) is not fulfilled, then B_2 becomes a different function of θ and the minimizing process is altered; an experimental plot of σ against ϱ might then be expected to conform to a curve which does not belong to the family in Fig. 8.11.

Obviously, Fig. 8.11 can only be used with confidence when its validity has been established by compressive tests on short sandwich columns of construction similar to those to which it is proposed to apply the analysis.

8.8. Some Special Observations on Wrinkling Behaviour

The effect of initial irregularity of the faces when the core is orthotropic was investigated by Norris.[6.3] The theory is too extensive to be quoted here, but it represents a generalization of the analysis for isotropic cores given in the preceding section. The results are presented in a graph similar in form to that in Fig. 8.11, using the same parameter k to describe the amplitude of the initial deformation.

The character of wrinkling behaviour is much the same whether the core is orthotropic or isotropic, with the following exceptions.

When the modulus of elasticity of the core normal to the faces is low in comparison with the shear modulus in the zx-plane, symmetrical wrinkling (Fig. 8.4, III) is more likely than antisymmetrical wrinkling. The possibility of symmetrical wrinkling cannot be disregarded as it can for isotropic cores. An orthotropic core of this special type is not likely to be used in practice because of its poor local resistance to pressures applied to the faces.

Symmetrical wrinkling is also theoretically possible when the modulus of elasticity of the core in the direction of the load (parallel with the faces) is exceptionally small compared with the modulus in the direction perpendicular to the faces. The physical significance of this situation is far from clear. However, a honeycomb core fulfills these conditions and an investigation of symmetrical wrinkling of sandwiches with honeycomb cores was made by Norris.[6.4] The results of the study were not conclusive and indeed the theory as a whole does not appear to be strictly applicable to honeycomb cores unless it can be shown that the cell size is much less than the wavelength of the wrinkles in the faces.

In this chapter, attention has been confined to the plane stress problem, in which $\sigma_y = 0$. This is legitimate provided the structure is narrow, as it might be in a sandwich beam or strut. When the structure is wide, as it usually is in a rectangular panel supported on four sides, it is commonly supposed that a condition of plane strain is appropriate, $e_y = 0$. Norris' analysis for orthotropic cores (previously mentioned) is applicable to either condition. For isotropic cores, the difference between plane stress and plane strain conditions is not large; it may be reduced but not eliminated by writing $E_f/(1 - v_f^2)$ in place of E_f.

The discussion of wrinkling in this chapter has been confined to the two-dimensional problem of a sandwich strut. The three-dimensional problem of wrinkling in a rectangular panel supported on four sides could be treated in an analogous manner. However, if the edge load is applied to the panel in the x-direction buckling usually occurs with only one half-wave from edge to edge in the y-direction. The wrinkling wavelength in the x-direction is likely to be much smaller. The variation of stresses in the y-direction in the core and faces is much less severe than the variation in the x-direction and it is reasonable to suppose that the situation does not differ materially from that of the two-dimensional problem; the same formulae may therefore be used in both sets of circumstances.

8.9. Local Instability of the Elements of a Sandwich, other than Wrinkling

Apart from the short-wavelength wrinkling instability of a face which is supported by a continuous elastic medium, various forms of local instability are possible where components of the sandwich are not continuously supported.

For example, a sandwich with a corrugated core (Fig. 7.3)

subjected to compression in the direction of the corrugations may fail because of simple plate-type instability of elements such as *ab* or *bb*. As a first approximation the element *bb* may be considered as a long narrow simply-supported rectangular plate; alternatively the interaction of the various plate elements in the cross-section may be investigated by standard methods.

A sandwich with a honeycomb core may fail by buckling of the face in the small region where it is unsupported by the walls of the honeycomb. This is sometimes termed intra-cellular bucking and various formulae have been proposed to describe it. The formula

$$\sigma_{cr} = \frac{E}{3}\left(\frac{t}{R}\right)^{\frac{3}{2}} \quad (8.35a)$$

was proposed by Norris,[6.6] where E is the modulus of elasticity of the face, t is the thickness of the face, R is the radius of the circle which can be inscribed in the honeycomb cell and σ_{cr} is the critical stress in the face. The formula was related to tests on faces attached to continuous cores with isolated holes drilled to represent the openings of a honeycomb core. In a later report by Kuenzi[16.3] the formula was verified by tests on some honeycomb sandwiches.

The formula (8.35a) may be questioned on theoretical grounds. If the shape of the cells is constant, a consideration of ordinary plate buckling formulae suggests that the equation should be of the form

$$\sigma = \text{constant} \times \left(\frac{t}{R}\right)^2. \quad (8.35b)$$

Such a formula, with the constant equal to $2E_f/(1-v_f^2)$, is proposed in ref. 5.12 (part III, chapter 5).

It is fairly certain that local instability depends to some extent on methods of fabrication. For example, the behaviour of the sandwich in Fig. 7.3 will depend on whether the corners of the core are sharp or rounded, and whether the interface is continuously bonded or discontinuously riveted or welded. Intracellular buckling in a honeycomb sandwich may be less severe if very thick glue lines are used. As a result, it is prudent to regard formulae such as (8.35) as a very rough guide at best, until they have been substantiated by tests on particular forms of construction.

8.10. Interaction of Wrinkling and Overall Instability

In the wrinkling analysis in Section 8.2 the faces were assumed to be thin and the possibility of displacement of the faces in the x-direction was excluded. Both of these limitations were removed in a more elaborate theory due to Williams, Leggett and Hopkins[8.1] Because the elaborate theory permits longitudinal displacement of the faces, it includes overall Euler-type instability as a special case. In fact, it provides a link between such overall instability on the one hand and wrinkling behaviour on the other. Only an outline of the theory, as applied to sandwiches with isotropic cores, will be given here.

The origin of the coordinates will be taken at the centre of the core in this instance. Suitable stress functions for the core may be written in the form:

$$\phi = (Fz^2/2) + \left(Az \sinh \frac{\pi z}{l} + D \cosh \frac{\pi z}{l}\right) \sin \frac{\pi x}{l}, \quad (8.36a)$$

$$\phi = (Fz^2/2) - \left(Bz \cosh \frac{\pi z}{l} + C \sinh \frac{\pi z}{l}\right) \sin \frac{\pi x}{l}, \quad (8.36b)$$

where A, B, C, D, F are constants and l is the half-wavelength of the sinusoidal displacement of the faces. Expressions (8.36a) and (8.36b) are appropriate for symmetrical and antisymmetrical displacements respectively (cases III and II, Fig. 8.4).

The stresses and strains in the core may be found by inserting the stress function in equations (8.14) and (8.15). The displacements in the core may be found by suitable integration of the strains. The conditions at the lower interface (suffix 2) are of interest; if z is equated to $+c/2$, the following results can be obtained for antisymmetrical displacements:

$$\sigma_{z2} = A_1 \frac{\pi^2}{l^2} \sin \frac{\pi x}{l}, \tag{8.37a}$$

$$\tau_{zx2} = A_2 \frac{\pi}{l} \cos \frac{\pi x}{l}, \tag{8.37b}$$

$$u_2 = \frac{1}{E_c} \left\{ Fx + A_3 \cos \frac{\pi x}{l} \right\}, \tag{8.37c}$$

$$w_2 = \frac{1}{E_c} \left\{ -(cFv_c/2) + A_4 \sin \frac{\pi x}{l} \right\}, \tag{8.37d}$$

where

$$A_1 = B \frac{c}{2} \cosh \frac{\pi c}{2l} + C \sinh \frac{\pi c}{2l}, \tag{8.38a}$$

$$A_2 = \left(B + C \frac{\pi}{l} \right) \cosh \frac{\pi c}{2l} + B \frac{\pi c}{2l} \sinh \frac{\pi c}{2l}, \tag{8.38b}$$

$$A_3 = \left[2B + C(1+v_c) \frac{\pi}{l} \right] \sinh \frac{\pi c}{2l} + B(1+v_c) \frac{\pi c}{2l} \cosh \frac{\pi c}{2l}, \tag{8.38c}$$

$$A_4 = B(1+v_c) \frac{\pi c}{2l} \sinh \frac{\pi c}{2l} + \left[C(1+v_c) \frac{\pi}{l} - B(1-v_c) \right] \cosh \frac{\pi c}{2l}. \tag{8.38d}$$

The sense of the stresses σ_{z2} and τ_{zx2} is shown in Fig. 8.12a; the displacements u_2, w_2 are positive when in the x-and z-directions. The displacements at the centre line of the face are:

$$w = w_2, \tag{8.39}$$

$$u = u_2 - \frac{t}{2} \frac{dw}{dx}. \tag{8.40}$$

Figure 8.12b shows a short length dx of the lower face together with the forces which act upon it. By taking moments about the centre it is easy to show that the transverse shear stress τ is equal to $\tau_{zx2}/2$. The net downward force on the element is $\left(bt(d\tau/dx) - b\sigma_{z2}\right)dx$, which is equivalent to a distributed vertical load of $(bt/2)(d\tau_{zx2}/dx) - b\sigma_{z2}$. The differential equation of cylindrical bending for a strip of lower face of unit width is therefore

$$\frac{E_f t^3}{12g} \frac{d^4w}{dx^4} + \sigma t \frac{d^2w}{dx^2} = \frac{t}{2} \frac{d\tau_{zx2}}{dx} - \sigma_{z2}. \qquad (8.41)$$

In this equation σ is the compressive stress in the faces and g is

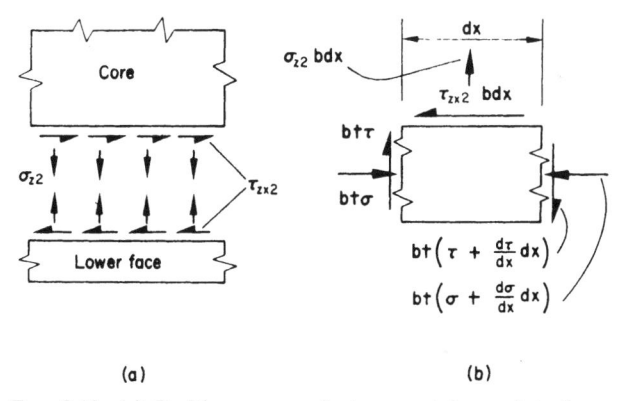

FIG. 8.12. (a) Positive senses of stresses at lower interface. (b) Forces acting on an element of a thick lower face.

equal to $(1 - v_f^2)$. The next equation expresses the horizontal equilibrium of the same element:

$$-b\tau_{zx2}\,dx = bt\frac{d\sigma}{dx}\,dx. \qquad (8.42)$$

With the aid of equation (8.40), the compressive stress in the face can be related to the displacements u_2 and w:

$$-\sigma = \frac{E_f}{g_1}\frac{du}{dx} = \frac{E_f}{g_1}\left(\frac{du_2}{dx} - \frac{t}{2}\frac{d^2w}{dx^2}\right). \qquad (8.43)$$

The term $g_1 = 1 - v_f^2$ is necessary if the face is in a state of plane strain; a distinction has been made between g and g_1 for the convenience of later remarks. Substitution for σ from (8.43) in (8.42) provides the following result:

$$\tau_{zx2} = t\,\frac{E_f}{g_1}\left(\frac{d^2u_2}{dx^2} - \frac{t}{2}\,\frac{d^3w}{dx^3}\right). \tag{8.44}$$

Equations (8.41) and (8.44) describe the equilibrium of the face in the z- and x-directions respectively. Substitutions may be made for $w(=w_2)$, u_2, σ_{z2} and τ_{zx2} from equations (8.37).

$$\frac{A_4}{E_c}\left\{\sigma t - \frac{E_f t^3}{12g}\,\frac{\pi^2}{l^2}\right\} = A_1 + tA_2/2, \tag{8.45}$$

$$A_2 + \frac{t\pi}{g_1 l}\,\frac{E_f}{E_c}\left\{A_3 - A_4\,\frac{t}{2}\,\frac{\pi}{l}\right\} = 0. \tag{8.46}$$

These are two homogeneous equations in the unknown constants B and C, as may be seen by using equations (8.38) to eliminate $A_1 - A_4$. One possible solution is $B = C = 0$, in which case there are no displacements at all and the face remains straight and unbuckled. A solution is possible when B and C are non-zero provided the determinant of the coefficients of B and C in equations (8.45) and (8.46) vanishes; such non-zero values of B and C indicate that the face has buckled. Some of the coefficients contain the stress σ; the determinantal equation can therefore be solved for the critical value of σ at which buckling takes place. The procedure is lengthy but straightforward and the result may be written in the following manner:

$$\frac{\sigma}{E_f} = \frac{F + GT + HT^2}{A + BT + CT^2}, \tag{8.47}$$

where

$A = \pi^3(1+\nu_c)^2 p^3 q^2 r/2 + 2\pi pqg_1,$

$B = \pi^2(3-\nu_c)(1+\nu_c)p^2 q^2 r,$

$C = -\pi^3(1+\nu_c)^2 p^3 q^2 r/2,$

$F = \pi^3 p^3 q^3 g_1/6g - \pi pg_1/2r + \pi^3 p^3 q^2(1+\nu_c+q)/2$
$\quad + \pi^5(1+\nu_c)^2 p^5 q^4 r/24g,$

$$G = g_1/r + \pi^2(1-\nu_c)p^2q^2 + \pi^4(1+\nu_c)(3-\nu_c)p^4q^4r/12g,$$
$$H = \pi p g_1/2r + 2\pi pq - \pi^3(1+\nu_c)p^3q^2/2 - \pi^5 p^5 q^4 r(1+\nu_c)^2/24g,$$
$$p = c/l, \quad q = t/c, \quad r = E_f/E_c, \quad T = \tanh\frac{\pi c}{2l}.$$

(This is identical with a result quoted in ref. 8.1 provided g_1 (but not g) is made equal to unity; in that case the faces bend locally in plane strain, but they extend in plane stress, while the core is in a state of plane stress also. If g are g_1 and both given the value $(1-\nu_f^2)$ and ν_c is made zero, equation (8.47) corresponds with a result in ref. 9.3, which is intended for use when faces and core are both in a state of plane strain.)

Equation (8.47) has been used in the preparation of Fig. 8.13 for a sandwich with rather thick faces ($t/c = 0\cdot 2$) and for $\nu_f = \frac{1}{3}$, $\nu_c = 0$, $g = g_1 = (1-\nu_f^2)$. The ordinate σ_{cr}/E_f is a non-dimensional measure of the buckling stress and the abscissa is marked in terms of l/d rather than l/c, where $d = c+t$.

The four curves (for different values of the ratio E_f/E_c) evidently relate the buckling stress to the length of the strut. The straight line aa represents the Euler stress for the faces when they buckle as two independent struts:

$$bt\sigma_{cr} = \frac{\pi^2 E_f}{gl^2} \cdot \frac{bt^3}{12}$$

or

$$\frac{\sigma_{cr}}{E_f} = \frac{\pi^2}{12g}\left(\frac{l}{d}\right)^{-2}\left(\frac{d}{t}\right)^{-2}. \tag{8.48}$$

The straight line bb represents the stress in the faces when the strut buckles as a composite structure without shear strain in the core:

$$2bt\sigma_{cr} = \frac{\pi^2 E_f}{g_1 l^2} \cdot \frac{btd^2}{2},$$

or

$$\sigma_{cr} = \frac{\pi^2}{4g_1}\left(\frac{l}{d}\right)^{-2}. \tag{8.49}$$

Except for the fact that Fig. 8.13 is plotted to logarithmic scales for convenience, it is exactly equivalent to Fig. 3.2. For example, the curves in Fig. 8.13 correspond to the strut curve ABC

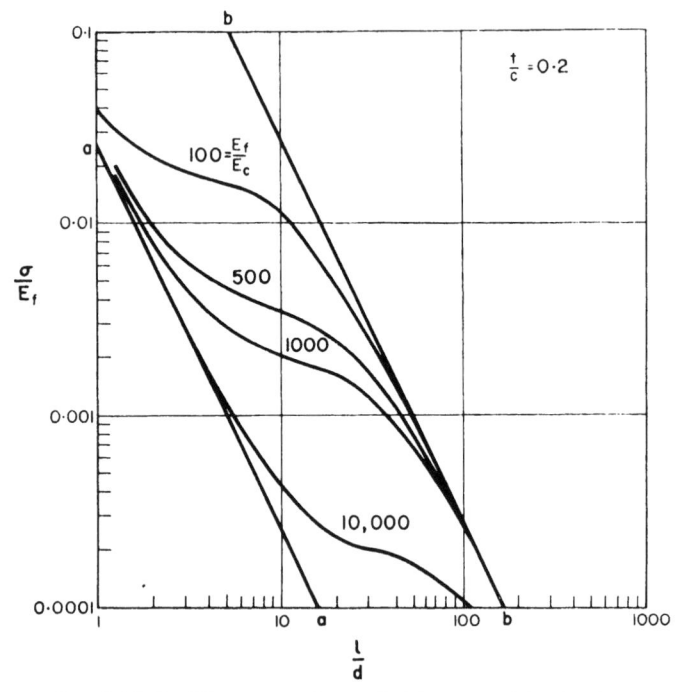

FIG. 8.13. Buckling curves for sandwich struts with thick faces, showing absence of wrinkling. ($v_c = 0$, $v_f = \frac{1}{3}$, cylindrical bending.) (Based on fig. 4 of ref. 9.3. By permission of the American Institute of Aeronautics and Astronautics.)

in Fig. 3.2, and the Euler lines aa, bb in Fig. 8.13 correspond to the Euler curves AFG and DEC respectively in Fig. 3.2.

So far equation (8.47) has provided no new information about the buckling behaviour of a sandwich strut. But in Fig. 8.14 a fresh set of curves is plotted, this time for a sandwich with rather thin faces ($t/c = 0.02$). Consider, for example, the curve for

$E_f/E_c = 1000$. A fairly short strut ($l/d = 3.5$) apparently buckles at a stress $\sigma_{cr} = 0.013 E_f$ (point c), but the presence of a minimum point (d) indicates that wrinkling will occur instead at a

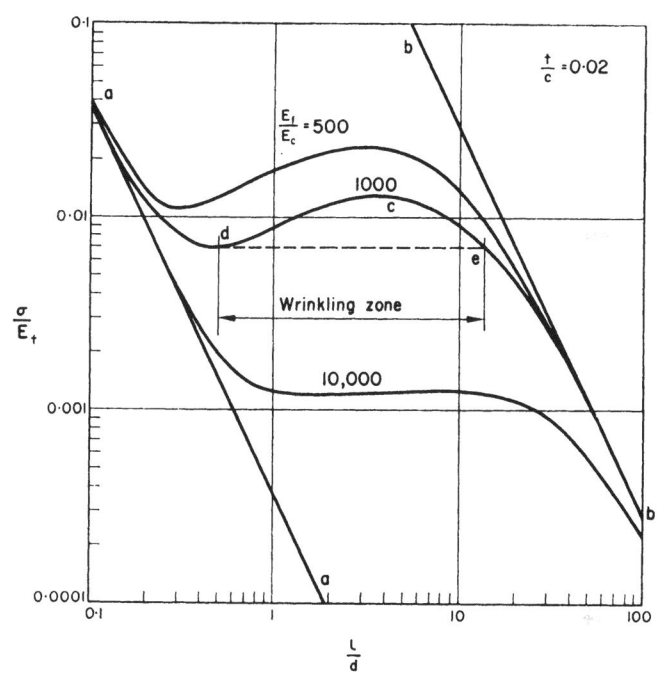

FIG. 8.14. Buckling curves for sandwich struts with thin faces, showing the possibility of wrinkling when $E_f/E_c = 500$ or 1000. ($v_c = 0$, $v_f = \frac{1}{3}$, cylindrical bending.) (Based on fig. 4 of ref. 9.3. By permission of the American Institute of Aeronautics and Astronautics.)

lower stress ($0.007 E_f$) and at a half-wavelength given by $l/d = 0.5$. In fact wrinkling at that particular half-wavelength will occur in all struts which fall in the region *dce*. Longer struts fail by overall Euler-type instability (modified by the shear deformation of the core); shorter struts (if they could be built) would fail by Euler instability of the faces acting as separate struts.

Equation (8.47) therefore describes the complete buckling behaviour of a sandwich strut, including the possibility of wrinkling. The accuracy of the equation is verified by Goodier[9.3, 10.2] in an even more thorough analysis of the problem. Williams, Leggett and Hopkins derive an analogous equation for symmetrical deformations, but they conclude that symmetrical instability always occurs at a higher stress than antisymmetric instability.

Although equation (8.47) provides a useful picture of the nature of wrinkling and its relationship with other kinds of instability, it is not suitable for routine use and, for practical purposes, the ordinary wrinkling theory of Sections 8.1–8.3 is quite adequate.

A comparison between the two analyses can be made if it is remembered that Sections 8.1-8.3 refer to anticlastic bending. To change to cylindrical bending it is merely necessary to replace E_f by $E_f/(1-v_f^2)$ in equations (8.11) and (8.12) and in Fig. 8.6. With this in mind, Fig. 8.6 shows that antisymmetrical wrinkling can occur only if:

$$\varrho = \frac{t}{c}\left\{\frac{E_f}{E_c(1-v_f^2)}\right\}^{\frac{1}{3}} < 0.5 \quad \text{for} \quad v_c = 0 \qquad (8.50a)$$

or

$$\frac{E_f}{E_c} < 13.9 \quad \text{for} \quad t/c = 0.2, \quad v_f = \frac{1}{3}, \qquad (8.50b)$$

$$\frac{E_f}{E_c} < 13{,}900 \text{ for} \quad t/c = 0.02, \quad v_f = \frac{1}{3}. \qquad (8.50c)$$

Condition (8.50b) implies that wrinkling is not to be expected at all in Fig. 8.13. Condition (8.50c) implies that wrinkling will occur for all the ratios E_f/E_c illustrated in Fig. 8.14. These implications are confirmed by the presence of minima in Fig. 8.14 and their absence in Fig. 8.13.

Furthermore, Table 8.3 lists the wrinkling stresses determined from (a) Fig. 8.6 for $v_c = 0$, $v_f = \frac{1}{3}$, and (b) the minimum point such as d in Fig. 8.14.

TABLE 8.3. COMPARISON OF FIGS. 8.6 AND 8.14

$\dfrac{E_f}{E_c}$	ϱ	Fig. 8.6			Fig. 8.14	
		B_1	$\dfrac{\sigma_{cr}}{E_f}$	$\dfrac{l}{c}$	$\dfrac{\sigma_{cr}}{E_f}$	$\dfrac{l}{d} \div \dfrac{l}{c}$
500	0·165	0·63	0·0104	0·35	0·011	0·3
1,000	0·208	0·63	0·0066	0·4	0·007	0·5
10,000	0·448	0·54	0·0012	1·5	0·0012	1·5

The two sets of values for σ_{cr}/E_f and l/c are as close as may be expected from the graphs available.

CHAPTER 9

THE DEVELOPMENT OF THE THEORY OF SANDWICH PANELS

9.1. Historical Development of Sandwich Theory

Very few papers have been published which deal with the bending and buckling of sandwich panels with cores which are rigid enough to make a significant contribution to the bending stiffness of the panel, yet flexible enough to permit significant shear deformations. The complexity of a problem of that kind is sufficient to place it beyond the scope of this book, except in the special cases mentioned briefly in Section 2.10. There remains the considerable problem of the sandwich panel with an antiplane core, one which possesses no stiffness in the xy-plane and in which the shear stresses τ_{zx}, τ_{yz} are constant throughout the depth (i. e. they are independent of z). Such panels differ from ordinary homogeneous plates in that the bending deformations may be enhanced by the existence of non-zero shear strains (γ_{zx}, γ_{yz}) in the core and of direct strains e_z in the core, perpendicular to the faces. The shear strain and the direct strain in the core are also directly associated with the possibility of short wavelength instability of the faces (wrinkling).

This problem has been the subject of two main methods of attack, which may be referred to for convenience as the general and the selective methods. In the general method equations are set up to define the equilibrium of the separate faces and of the core and to prescribe the necessary continuity between the faces and the core. The result is a set of differential equations which may be solved in particular cases for the transverse deformations of the panel, the flattening of the core and other quantities of interest. In the selective method, which has been the basis of this book, the problem is divided into two separate parts which may be named (again for convenience) as the bending problem and the wrinkling problem. In the bending problem it is convenient to assume that the core is not only antiplane, but also infinitely stiff in the z-direction. This excludes the possibility of flattening of the core and of wrinkling instability, but it does permit the assessment of the effect of core shear deformation on the deflections and stresses in the panel. In the wrinkling problem the true elastic properties of the core are taken into account but the task is simplified by permitting the middle planes of the faces to deflect in the z-direction only, not in their own planes. In this way overall bending of the panel is excluded, but the phenomena of wrinkling and of local distortion under concentrated load can be studied.

The general method

The general method was investigated by Reissner[4.1] (1948) in relation to isotropic panels with very thin faces. Although his analysis is not simple, it is possible for Reissner to conclude that the effect of core flexibility in the z-direction is after all less important than effect of core shear deformation in the transverse planes. Wrinkling instability as such is not discussed. It is only by neglecting the effect of direct transverse core strains that Reissner is able to derive a relatively simple differential equation for the transverse displacement w. It is equivalent to equation (9.1) except that the in-plane forces N are expressed in terms of a stress function, and D_f is zero.

$$-\frac{D_f D_2}{D_Q} \nabla^6 w + (D_2 + D_f) \nabla^4 w = \left\{1 - \frac{D_2}{D_Q} \nabla^2\right\} \left\{q + N_x \frac{\partial^2 w}{\partial x^2} + 2N_{xy} \frac{\partial^2 w}{\partial x \, \partial y} + N_y \frac{\partial^2 w}{\partial y^2}\right\}, \quad (9.1)$$

where $\nabla^2 = \dfrac{\partial^2}{\partial x^2} + \dfrac{\partial^2}{\partial y^2};\quad \nabla^4 = \nabla^2\nabla^2;\quad \nabla^6 = \nabla^2\nabla^2\nabla^2;$

$$D_2 = \frac{E_f t d^2}{2(1-\nu_f^2)};\quad D_Q{}^\dagger = G_c d^2/c;\quad D_f = E_f t^3/6,$$

and q, N_x, N_y, N_{xy} are functions of x and y.

An equation very similar to (9.1) is obtained by Eringen[5.5] (1951) who neglects the geometrical thickness of the (equal) faces but includes their local bending stiffnesses and also the bending stiffness of the core. However, the inclusion of the latter is contradicted to some extent by the assumption that the u and v displacements in the core are directly proportional to z.

A much more recent analysis by Heath[5.13] (1960) also includes a very similar equation, but for a sandwich with an orthotropic core. Heath's analysis is based on earlier work by Hemp [3.4] (1948) and is apparently independent of the work of Reissner.

Raville[1.13] (1955) applies the general method to the problem of a simply-supported rectangular panel with uniform transverse load and with thin faces. The three displacements of points in the orthotropic antiplane core are expressed as polynomials in z, but the complexity of the analysis again makes it necessary to revert to the simplifying assumption of infinite core stiffness in the z-direction.

For practical purposes the general method is evidently intractable when applied to sandwich panels, but more success has been achieved in relation to sandwich struts and beams. The early work of Williams, Leggett and Hopkins[8.1] (1941) and of Cox and Riddell[8.2] (1945) falls into this category. The first of these deals with a sandwich strut with thick faces and an isotropic core (with an extension for orthotropic cores) and the analysis is used to form a link between the extreme cases of wrinkling instability (no longitudinal displacements of the faces during buckling) and

---

† Written as G_c/c in the original paper, which neglects the thickness of the faces.

of overall Euler-type instability, modified for shear deformations in the core (no direct core strains in the z-direction). A summary of the method is given in Section 8.10.

A very thorough analysis of the behaviour of struts with isotropic faces and cores has been outlined by Goodier[10.2] (1946) and completed by Goodier and Neou[9.3] (1951). In the latter paper the works of Williams[8.1] and of Cox[8.2] are verified to a high degree of accuracy.

Selective method; bending problem

Most of the published work on sandwich panels refers to the selective method and, in particular, to the bending problem, in which core strains in the z-direction are neglected. The assumption that the core is weak in the xy-plane leads in any case to the conclusions that the core makes no contribution to the flexural rigidity of the sandwich, that the core shear stresses τ_{zx} and τ_{yz} are independent of z and that a straight line drawn in the unloaded core normal to the faces remains straight after deformation, but is no longer normal to the faces. These assumptions (core weak in xy-plane, stiff in z-direction) allow the displacements of the panel to be expressed in terms of only three variables, one of which is the transverse displacement w. The other two variables are a matter of choice; some of the variables used by different authors are listed in Table 9.1.

The theory appears to have been developed in several equivalent but largely independent streams. The first stream originated with Williams, Leggett and Hopkins[8.1] (1941) who use w and λ (Table 9.1) to define the displacements of a sandwich strut. The same authors (1941,[8.1] 1942,[3.1] 1944[3.2]) use the variables w and $\lambda = \mu$ in a series of reports on the buckling and bending of isotropic sandwich panels. March[1.5] (1948) uses the same procedure for orthotropic panels. In all these reports it is assumed that λ and μ are equal and constant; it has been shown in Chapter 7 of this book that λ and μ are indeed constant under certain

Table 9.1. Variables Used by Different Authors to Define the Deformations of a Sandwich Plate

In every case the third variable is *w*. All variables are functions of *x* and *y* in general.

HOFF[(2.5)] (Upper and lower faces equal)	u, v = Displacements of middle plane of upper face in x- and y-directions. $-u, -v$ = Corresponding displacements of lower face. For equal faces only, $u = \left(\dfrac{\lambda c + t}{2}\right) \dfrac{\partial w}{\partial x}; \quad v = \left(\dfrac{\mu c + t}{2}\right) \dfrac{\partial w}{\partial y}.$
U.S. FOREST PRODUCTS LABORATORY	λ and μ, where $\gamma_{zx} = (1 - \lambda) \dfrac{\partial w}{\partial x},$ $\gamma_{yz} = (1 - \mu) \dfrac{\partial w}{\partial y}.$
REISSNER[(4.1)]	α, β = rotations in zx- and yz-planes of the normal in the core. $\alpha = \lambda \dfrac{\partial w}{\partial x}; \quad \beta = \mu \dfrac{\partial w}{\partial y}.$
LIBOVE and BATDORF[(2.1)]	Q_x, Q_y Or γ_{zx}, γ_{yz} where: $Q_x = G_{zx} \dfrac{d^2}{c} \gamma_{zx}; \quad Q_y = G_{yz} \dfrac{d^2}{c} \gamma_{yz}.$

conditions and that they can be equal only when the sandwich is isotropic.

This last fact was soon recognized in very comprehensive analyses by Ericksen and March.[(1.7)] In a report of 1950 (revised 1958) these authors determine the critical edge loads (N_x) of sandwich plates with various boundary conditions. Essentially the technique is to assume that λ and μ are different unknown constants and to minimize the strain energy with respect to λ, μ and the unknown amplitudes of the assumed deflection modes (Rayleigh–Ritz method). It is not certain that all the boundary conditions are fulfilled by the chosen functions, nor that λ and μ are indeed constants in every case considered; to that extent the critical loads are approximate (except for the simply-supported case) and have been considered from 5 to 10% high (see Thurston[4.5]). However, the results are of very wide application and can be used for sandwiches with thick faces of unequal thickness and different orthotropic materials and orthotropic cores. Ericksen[1.8, 1.9] (1950, 1951) uses a similar technique for the bending of simply-supported panels, except that a modification is made to the Rayleigh–Ritz procedure to improve the convergence of the series for deflections and stresses. This type of analysis has been the basis of a number of reports by the U.S. Forest Products Laboratory,[1.4, 1.11, 1.12, 1.15, 1.16, 1.17, 1.18] which has also been responsible for extensive tests on sandwich panels.[1.2, 1.3, 1.6, 1.10, 1.12] Chapters 4, 5 and 6 of this book are based on this approach.

A second stream of development flows from the work of Libove and Batdorf[2.1] (1948, 1950) who derive three differential equations for the three unknowns w, Q_x and Q_y. The properties of the sandwich panel are expressed in terms of flexural, torsional and shear rigidities rather than in terms of the dimensions and material properties of the individual components of the sandwich. The analysis is valid for orthotropic panels with *very thin* or *thin* faces of different thicknesses and materials. An outline of the method is given at the beginning of Chapter 7, which also indicates the form in which the differential equations can be obtained ((7.13), (7.20); and (7.21), or (7.22a, b, c)). In general it is not possible to assume any particular relationship between w, Q_x and Q_y, other than that implied by the three differential equations. However, it happens that is the case of a simply-supported plate with sinusoidal transverse load and edge loads N_x, N_y the three variables are connected by the equations (7.24a, b) in which λ' and μ' are constants. When the panel is isotropic it is not difficult to show that, apart from extra differential operators common to both sides, equation (7.22a) (the differential equation for w, derived by Libove and Batdorf[2.1]) coincides with Reissner's[4.1] equation ((9.1) with $D_f = 0$). Libove and Batdorf also derive an expression for the strain energy in an orthotropic plate, again in terms of w, Q_x, Q_y and their derivatives, which can be used for solutions of the Rayleigh–Ritz type. An invaluable adjunct to the paper by Libove and Batdorf is another one by Libove and Hubka[12.4] (1951) which shows in considerable detail how the various stiffnesses of a corrugated-core sandwich may be found.

The equations of Libove and Batdorf have found practical application in a paper by Seide and Stowell[2.2] (1949), who investigate the instability of an isotropic simply-supported panel with very thin faces by direct substitution of sinusoidal deflection functions in the differential equation for w (equations (9.1) or (7.22a)). Seide[2.3, 2.9] (1952) has also investigated buckling with simply-supported loaded edges and clamped unloaded edges. This time the variables w, Q_x and Q_y are expressed as products of sine (or cosine) functions of x and the sum of hyperbolic functions of y. These are inserted in the differential equations of Libove and Batdorf and the hyperbolic functions are adjusted to satisfy the clamped boundary conditions. Seide[2.4] (1949) has also investigated buckling due to edgewise shear forces (N_{xy}).

The equations of Libove and Batdorf have also been put to practical use by Robinson[5.7] (1955), who analyses the bending and buckling of simply-supported panels by assuming sine and cosine functions for w, Q_x and Q_y. The results happen to be exact, because such deflection functions satisfy the differential equations and the boundary conditions; they are also closely analogous with the results of the later part of Chapter 7 of this book. Robinson's results are applicable to panels with isotropic faces and orthotropic cores of honeycomb or corrugated types; they are valid for thin unequal faces.

A further extension of this work has been made by Harris and Auelmann[4.6] (1960) who use the strain energy expression of Libove and Batdorf to study the instability of simply-supported corrugated-core panels. The core shear stiffness is taken as infinite in the zx-plane; in the yz-plane a constant ("g") is introduced which is equivalent to μ' in equation (7.24b) of this book. The strain energy is then minimized with respect to the unknown amplitudes of sinusoidal deflection functions (Rayleigh–Ritz). Results are given for many combinations of the edge loads N_x, N_y, N_{xy}. The analysis is restricted to sandwiches with very thin isotropic faces and with corrugated cores which make a negligible contribution to the flexural rigidity of the panel; in a secondary study it is shown that the magnitude of the flexural rigidity of the core has only a minor influence on the buckling load. It seems likely that the results for uniform edge loads N_x, N_y are exact, whereas those for in-plane bending and for N_{xy} are approximate because the chosen sinusoidal functions satisfy the differential equations only in the former case (see Section 7.7).

A third stream of development appears to have originated with Hoff[2.5] (1950). In his paper the strain energy of a sandwich plate in bending is expressed in terms of the transverse displacement (w) and the in-plane displacements u and v of the middleplane of the upper face (Table 9.1). The in-plane displacements of the lower face are $-u$ and $-v$. A variational method is used to derive three differential equations for u, v and w which express the condition that the potential energy shall be a minimum. Finally, u and v are eliminated to leave a single differential equation for w, identical with equation (9.1) except for the omission of N_{xy} Hoff's equation is therefore a generalization of Reissner's, and takes into account the flexural rigidity D_f of the faces. It is shown in Appendix II that equation (9.1) is the analogue of the differential equation for a beam column which is implicit in Chapters 2 and 3.

Hoff's equations have been used by Yen, Salerno and Hoff[2.6] (1952) to establish upper and lower bounds to the buckling load N_x of an isotropic sandwich plate with clamped loaded edges and simply-supported unloaded edges. Nardo[4.4] (1953) considers the same problem and also that of a plate with all edges clamped. Yen, Gunturkin and Pohle[2.7] (1951) use sinusoidal functions in Hoff's equation for w in order to estimate the deflection of an isotropic simply-supported plate with uniform transverse load and with a central point load. The effect of face thickness is indicated.

In a paper by Thurston[4.5] (1957), Hoff's energy expression is used in the buckling analysis of clamped plates. The total energy is minimized with respect to the unknown amplitudes of sine/cosine modes for u, v and w. The individual modes do not satisfy the boundary conditions, but the sums of the various modes are made to do so approximately. This imposes restrictions on the amplitudes, which are not all independent variables. Minimization is therefore achieved by the use of the Lagrangian multiplier method.

A comparison of the strain energy expressions used in the three streams of development just described is made in Section 9.2.

Standing somewhat apart from the evolution described above is a series of papers by Bijlaard[2.10, 4.2] which describe methods of predicting the buckling loads of sandwich structures from a knowledge of the buckling loads in the two (relatively simple) extreme cases in which the core is stiff and weak in shear. Equation (3.6b) affords a very simple illustration of the method.

Selective method; wrinkling problem

The literature of the wrinkling problem is less extensive than that of the bending problem.

The first major paper is by Gough, Elam and De Bruyne[10.1]

(1940); this paper contains an examination of the stability of a straight strut stabilized in various ways by an isotropic elastic medium. Some of the cases considered are directly applicable to the compression faces of sandwich beams and to the antisymmetrical wrinkling of sandwich struts; these have been incorporated in Chapter 8 of this book (cases I and II). An analysis of the same kind is made by Hoff and Mautner[9.1] (1945) for symmetrical wrinkling of sandwich struts; this appears as case III in Chapter 8. These authors also investigate a kind of antisymmetric buckling which they refer to as a "skew ripple". The deformation chosen was suggested by the buckled shape of a number of test specimens, but it is not certain whether or not this is really a degenerate form of the antisymmetric wrinkling discussed as case II in Chapter 8. The formula proposed by Hoff and Mautner for the "skew ripple" wrinkling stress is more complicated than the case II formula but the two formulae give results which are not very different is magnitude. The work of Williams, Leggett and Hopkins[8.1] (1941) in linking the wrinkling theory with the theory of instability of struts has already been mentioned and it is discussed briefly in Section 8.10. That of Cox and Riddell[8.2] (1945) represents an extension of the earlier work to cover the behaviour of sandwich members with orthotropic cores, subject to axial thrust and bending simultaneously. Experimental verifications of Cox's formulae are provided by Barwell and Riddell[8.3] (1946). Wan[9.2] (1947) appears to be the first to discuss the effect of initial irregularities on the wrinkling behaviour of the faces, especially in relation to the tensile strength of the adhesive. The analysis is analogous with that of Section 8.7 except that Wan is concerned only with symmetrical wrinkling and a core with rather special orthotropic properties intended to represent end-grain balsa. He also discusses wrinkling in a biaxial stress system.

The massive study by Norris et al.[6.3] (1949) is in effect an application of Wan's concept of initial irregularities to the wrinkling analysis of Gough, Elam and De Bruyne, with the added refinement of an orthotropic core. Section 8.7 is a special case of Norris's analysis, for isotropic cores only. In a later report, Norris[6.4] *et al.* (1953) extend the analysis further in order to study symmetric wrinkling in sandwiches with honeycomb cores; however, the validity of the application is questionable if the wrinkling wavelength is of the order of the cell size. The test results show that the buckling stress is largely independent of core depth, a conclusion which is compatible also with intracellular buckling. Norris and Kommers[6.6] (1950) and Kuenzi[16.3] (1951) report attempts to derive a formula for the stress at which intracellular buckling occurs.

Goodier and Neou[9.3] (1951) compare all the major wrinkling formulae available at that time and show them to be in fairly close agreement with their own wrinkling formula, derived from a most rigorous study of the general instability of a strut with an isotropic core. Goodier and Hsu[9.4] (1954) discuss the possibility of wrinkling at the ends of sandwich struts. This type of wrinkling is associated with non-sinusoidal deformations which decay at a distance from the ends of the strut. The analysis is for antiplane cores.

Later papers have been published by Yussuff[10.4, 10.5] (1955, 1960). Yussuff's analysis follows the same general principles as earlier authors, but he makes certain simplifying assumption (including the use of the Winkler hypothesis — see Section 8.5).

9.2. Comparison of Some Common Notations

Three of the main streams of development of the theory of bending of sandwich panels have already been referred to. In this section, the notations commonly used in each of the theories will be compared briefly. In this way it will be possible for the reader to make comparisons between the numerous papers which have been published, and to understand something of the different assumptions on which they are based. It is convenient to take the expressions for the strain energy in a bent sandwich panel as the basis of the comparison. Table 9.2 shows the strain energy expressions due to Libove and Batdorf[2.1] and to Hoff.[2.5]

TABLE 9.2

LIBOVE and BATDORF[2.1]

$$U = \frac{1}{2} \int_A \left\{ \frac{D_x}{g} \left[\frac{\partial}{\partial x} \left(\frac{\partial w}{\partial x} - \gamma_{zx} \right) \right]^2 + 2\nu_y \frac{D_x}{g} \frac{\partial}{\partial x} \left(\frac{\partial w}{\partial x} - \gamma_{zx} \right) \frac{\partial}{\partial y} \left(\frac{\partial w}{\partial y} - \gamma_{yz} \right) \right.$$

$$\left. + \frac{D_y}{g} \left[\frac{\partial}{\partial y} \left(\frac{\partial w}{\partial y} - \gamma_{yz} \right) \right]^2 + \frac{D_{xy}}{2} \left[\frac{\partial}{\partial x} \left(\frac{\partial w}{\partial y} - \gamma_{yz} \right) + \frac{\partial}{\partial y} \left(\frac{\partial w}{\partial x} - \gamma_{zx} \right) \right]^2 \right.$$

$$\left. + D_{Qx} \gamma_{zx}^2 + D_{Qy} \gamma_{yz}^2 \right\} dA.$$

HOFF[2.5]

$$U_{fm1} = U_{fm2} = \frac{Et}{2g} \int_A \left\{ \left(\frac{\partial u}{\partial x} \right)^2 + 2\nu \frac{\partial u}{\partial x} \frac{\partial v}{\partial y} + \left(\frac{\partial v}{\partial y} \right)^2 \right.$$

$$\left. + \frac{1-\nu}{2} \left[\frac{\partial u}{\partial y} + \frac{\partial v}{\partial x} \right]^2 \right\} dA,$$

$$U_{fb1} = U_{fb2} = \frac{Et^3}{24g} \int_A \left\{ \left[\frac{\partial^2 w}{\partial x^2} + \frac{\partial^2 w}{\partial y^2} \right]^2 \right.$$

$$\left. - 2(1-\nu) \left[\frac{\partial^2 w}{\partial x^2} \cdot \frac{\partial^2 w}{\partial y^2} - \left(\frac{\partial^2 w}{\partial x \partial y} \right)^2 \right] \right\} dA,$$

$$U_c = \frac{Gc}{2} \int_A \left\{ \left[\frac{2u}{c} - \frac{d}{c} \frac{\partial w}{\partial x} \right]^2 + \left[\frac{2v}{c} - \frac{d}{c} \frac{\partial w}{\partial y} \right]^2 \right\} dA.$$

Notes: $g = 1 - $ (product of Poisson's ratios),

u, v = displacements of middle plane of upper face,

U = strain energy of complete sandwich,

U_c = strain energy of core,

U_{fm1}, U_{fm2} = membrane strain energy of upper/lower face,

U_{fb1}, U_{fb2} = bending strain energy of upper/lower face.

The equation due to Libove and Batdorf is based on the arguments set out in Sections 7.1–7.3, and the flexural, twisting and shear stiffnesses $(D_x, D_y, D_{xy}, D_{Qx}, D_{Qy})$ are defined in Section 7.2. In the original form depicted in Table 9.2, the equation is valid only for sandwiches with very thin faces, although the faces may be of unequal thickness and different orthotropic materials (provided the principal axes of the materials coincide with the x- and y-directions). The equation can be made valid for thin faces (in which the geometrical effect of the face thickness is taken into account) by replacing γ_{zx} and γ_{yz} by $(c/d)\gamma_{zx}$ and $(c/d)\gamma_{yz}$. Evidently the equation expresses the strain energy in terms of $w, \gamma_{zx}, \gamma_{yz}$; in general these are three different functions of x and y. The Libove and Batdorf equation does not include the strain energy associated with the bending of the faces about their own axes. One important advantage of the equation is that it can be used without modification for corrugated-core sandwiches; the flexural rigidity of the core in such a sandwich is easily incorporated in D_x or D_y.

The equation due to Hoff is restricted to sandwiches with isotropic faces and cores. Because there is no stretching of the middle plane of the sandwich, the deformation is defined by the transverse displacement w, and the horizontal displacements u, v of the middle plane of the upper face. The horizontal displacements of the middle plane of the lower face are $-u, -v$. This symmetry implies that the equations are restricted to plates with identical upper and lower faces. The first and second expressions define the membrane energy and the local bending energy in each face; they are standard results. The third expression defines the shear strain energy in the core; it differs slightly from the original expression quoted by Hoff, and, unlike the latter, it makes full allowance for the thickness of the faces (see Fig. 9.1). The Hoff equations for $U_{fm1} + U_{fm2} + U_c$ may be derived from the Libove and Batdorf equations by the substitutions listed below; the expressions for U_{fb1} and U_{fb2} do not appear in the Libove and Batdorf equation. In general

THE THEORY OF SANDWICH PANELS 203

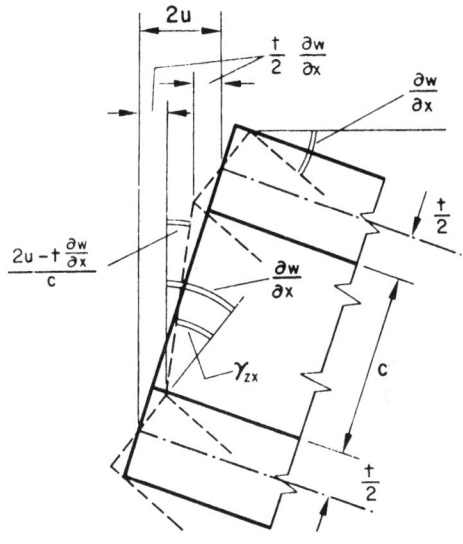

FIG. 9.1. Deformation in zx-plane of sandwich panel with thick equal faces.

$$\gamma_{zx} = \frac{\partial w}{\partial x} - \frac{1}{c}\left(2u - t\frac{\partial w}{\partial x}\right) = \frac{d}{c}\frac{\partial w}{\partial x} - \frac{2u}{c}.$$

u, v and w are all functions of x and y.

$$\left.\begin{array}{l} D_x = D_y = \dfrac{E_f t d^2}{2}; \quad D_{xy} = G_f t d^2 = \dfrac{E_f t d^2}{2(1+\nu_f)}; \\[6pt] D_{Qx} = D_{Qy} = G_c d^2/c. \\[4pt] \text{Replace } \gamma_{zx} \text{ and } \gamma_{yz} \text{ by } \dfrac{c}{d}\gamma_{zx} \text{ and } \dfrac{c}{d}\gamma_{yz} \text{ to allow for} \\[4pt] \text{thickness of faces.} \\[4pt] \gamma_{zx} = \dfrac{d}{c}\dfrac{\partial w}{\partial x} - \dfrac{2u}{c}, \quad \gamma_{yz} = \dfrac{d}{c}\dfrac{\partial w}{\partial y} - \dfrac{2v}{c}\text{ (Fig. 9.1).} \end{array}\right\} \quad (9.2)$$

The strain energy expression due to Ericksen and March[1.7] is not given in the table, but can be taken as the sum of equations (6.10), (6.11b), (6.12), (6.13) and (6.14). Here again the strain energy is a function of three quantities, w, λ and μ. In a general formulation of the equations these would all be functions of x and y. However, in the particular formulation given in Chapter 6 (also in Chapters 4 and 5) it has been assumed in the derivation of the strains from the displacements that λ and μ are constants. This means, for example, that the rotations of the normal in the core (which are defined as $\lambda(\partial w/\partial x)$ and $\mu(\partial w/\partial y)$ in the zx-and yz-planes) must be proportional to the slopes of the faces. Such a condition will only arise in a few problems, such as a simply-supported plate with sinusoidal transverse load and uniform edge loads N_x, N_y; in all other problems the equations represent an approximation to the truth. To compensate for this, the Ericksen and March equations are of great generality in that they are available for panels with orthotropic cores and with orthotropic faces of different thicknesses and materials. Even here, however, an approximation is involved if the faces are different in any way. Consider, for example, the first integral in equation (6.11b). The quantity q in the first term represents the distance of the centroid of the section, bending in the zx-plane, below the upper interface. It would usually be evaluated as:

$$q_1 = d\frac{E_{x2}t_2}{E_{x1}t_1+E_{x2}t_2} - \frac{t_1}{2}. \qquad (9.3a)$$

The analysis is strictly valid only if this coincides with the point B in Fig. 9.2, for which

$$q_2 = \frac{ct_1}{t_1+t_2}. \qquad (9.3b)$$

Furthermore, the quantity q in the second integral in equation (6.11b) represents the distance of the "centroid" of the section, in twisting, below the upper interface. It would usually be evaluated as:

$$q_3 = d\frac{G_{xy2}t_2}{G_{xy1}t_1+G_{xy2}t_2} - \frac{t_1}{2}. \qquad (9.3c)$$

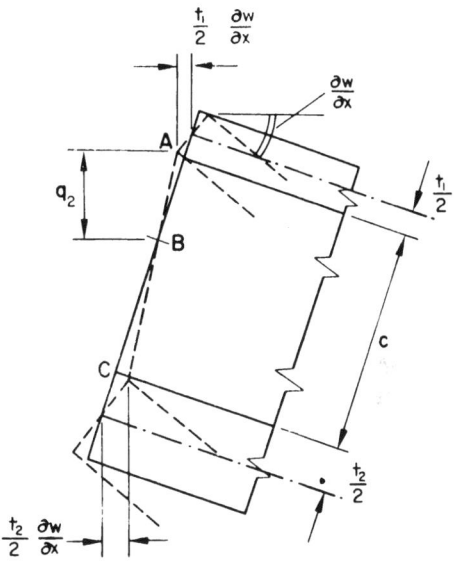

FIG. 9.2. Deformation in zx-plane of sandwich panel with thick unequal faces. (See equation (9.3b).)

For the purpose of equation (6.11b) one value of q has been used to represent a kind of weighted average of these three distinct values. Clearly, the result is approximate except in either of the following cases:

(i) The faces are identical but *thin*.

$$q_1 = q_2 = q_3 = \frac{c}{2}. \tag{9.4a}$$

(ii) The faces are *very thin* but different (subject to the limitation below).

If the faces are very thin, q_2 is no longer relevant and the quantity $t_1/2$ may be neglected in the expressions for q_1 and q_3. However, it is still necessary that

$$q_1 = q_3 \quad \text{and} \quad r_3 = r_1 \quad \text{or} \quad \frac{E_{x1}}{E_{x2}} = \frac{G_{xy1}}{G_{xy2}} = \frac{E_{y1}}{E_{y2}}. \tag{9.4b}$$

Fortunately, in practical sandwiches with faces of different materials, the divergence from the requirements of case (ii) is not likely to be very severe except in the most exotic combination of face materials.

It is possible to derive the Ericksen and March[1.7] equations from those of Libove and Batdorf[2.1] by the following substitutions:

Case (i)

$$D_x = \frac{E_x t d^2}{2}; \qquad D_y = \frac{E_y t d^2}{2}; \qquad D_{xy} = G_{xy} t d^2;$$

$$D_{Qx} = \frac{G_{zx} d^2}{c}; \qquad D_{Qy} = \frac{G_{yz} d^2}{c}.$$

Replace γ_{zx} and γ_{yz} by $\frac{c}{d}\gamma_{zx}$ and $\frac{c}{d}\gamma_{yz}$ to allow for thickness of faces. \hfill (9.5)

$$\gamma_{zx} = (1-\lambda)\frac{\partial w}{\partial x}; \qquad \gamma_{yz} = (1-\mu)\frac{\partial w}{\partial y};$$

λ, μ constant.

Case (ii)

$$D_x = E_{x1} t_1 q^2 + E_{x2} t_2 (d-q)^2;$$
$$D_y = E_{y1} t_1 q^2 + E_{y2} t_2 (d-q)^2,$$
$$D_{xy} = 2 d^2 G_{xy1} t_1 G_{xy2} t_2 / (G_{xy1} + G_{xy2}),$$
$$D_{Qx} = G_{zx} d; \qquad D_{Qy} = G_{zy} d; \qquad (9.6)$$
$$\gamma_{zx} = (1-\lambda)\frac{\partial w}{\partial x}; \qquad \gamma_{yz} = (1-\mu)\frac{\partial w}{\partial y};$$

λ, μ constant.

The assumption that λ and μ are independent of x is made only within each particular mode. Where the deformation of the panel is taken as the sum of a number of different modes, each with its own constant values for λ and μ, the overall deformation may be associated with a resultant pair of parameters, λ_r, μ_r, say, which define the total rotations of the core normals and which are functions of x and y. In such circumstances the limitations imposed on the Ericksen and March equations by the constancy of λ and μ are removed.

Neither the Hoff nor the Ericksen and March equations can cope with corrugated-core sandwiches without the introduction of special additional terms to represent the bending energy of the corrugated core.

9.3. Boundary Conditions

When the faces are thick the differential equation for w is of the sixth order, as in equation (9.1) This implies the existence of three boundary conditions for w at each edge.

For example, at the simply-supported edge $x = 0$ (Fig. 9.3a) the deflection w is zero, the local bending moments in the faces are zero and the overall bending moment (associated with membrane stresses in the faces) is also zero.

At the clamped edge $x = 0$ (Fig. 9.3b) the deflection w and the slope $\partial w/\partial x$ are both zero. Also, the line which joins the middle planes of the faces must not rotate in the zx-plane.

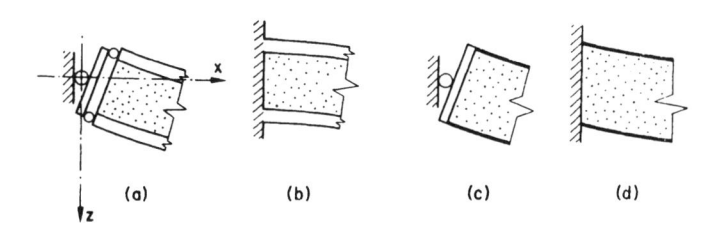

FIG. 9.3. Boundary conditions. (a) Thick faces, simple support. (b) Thick faces, clamped. (c) Thin faces, simple support. (d) Thin faces, clamped.

These boundary conditions are sufficient to determine the transverse displacement w. However, if the stresses in the plate are required it is also necessary to determine the other two unknowns in the problem (u and v, α and β, γ_{zx} and γ_{yz} or Q_x and Q_y according to choice). This involves further integrations and one extra boundary condition at each edge. At the edge $x = 0$, for example, it is usually convenient to choose *either* $\gamma_{yz} = 0$ *or* $M_{xy} = 0$. The condition $\gamma_{yz} = 0$ is equivalent to the insertion of an edge stiffener which is rigid in the yz-plane but free to twist. The condition $M_{xy} = 0$ implies that there is no shear stress τ_{xy} at the edges of the faces. The former condition is more realistic but the difference between the two is not likely to be significant unless the panel is loaded in a manner which tends to cause relative rigid-body movement of the faces in the xy-plane.

When the faces are very thin the term involving D_f in equation (9.1) vanishes, leaving a fourth-order equation. Only two boundary conditions are now required for w at each edge instead of three. At the simply-supported edge $x = 0$ (Fig. 9.3c) the deflection w is zero and the overall bending moment (associated with membrane stresses in the faces) is also zero.

At the clamped edge $x = 0$ (Fig. 9.3d) the deflection is zero and the line which joins the faces does not rotate in the zx-plane. The slope $\partial w/\partial x$ is *not* now zero.

In either case, the extra boundary condition $\gamma_{yz} = 0$ or $M_{xy} = 0$ is required if the stresses in the plate are to be evaluated.

9.4. Panels with Edges which are not Simply-supported

The solutions in Chapters 5, 6 and 7 are obtained by the use of sinusoidal deflection functions, which happen to satisfy the differential equations (see, for example, Section 7.7) and the boundary conditions for a simply-supported panel with edge members (Sections 5.6 and 9.3). Panels with other boundary conditions (e.g. with clamped edges) are less tractable because of the difficulty of finding suitable deflection functions. There are several methods of tackling such problems; three methods which have been used in the past are described briefly below.

(i) RAYLEIGH–RITZ METHOD

The procedure is to choose deflection functions which satisfy the boundary conditions but not the differential equations. Static equilibrium is then achieved by minimizing the strain energy of the system with respect to the unknown amplitudes of the chosen modes. Ericksen and March[1.7] use this procedure to investigate the buckling of panels under edge loads N_x; the deflection functions chosen for the various combinations of boundary conditions are as follows.

(a) Loaded edges simply supported, unloaded edges clamped

$$w = a_{m1} \sin \frac{m\pi x}{a} \sin^2 \frac{\pi y}{b} \qquad m = 1, 2, 3 \ldots \qquad (9.7)$$

(b) Loaded edges clamped, unloaded edges simply supported

$$w = a_{m1} \sin \frac{\pi x}{a} \sin \frac{m\pi x}{a} \sin \frac{\pi y}{b} \qquad m = 1, 2, 3 \ldots \quad (9.8)$$

(c) All edges clamped

$$w = a_{m1} \sin \frac{\pi x}{a} \sin \frac{m\pi x}{a} \sin^2 \frac{\pi y}{b}. \qquad (9.9)$$

Any of these functions may be used in the analysis of Chapter 6; the problem of orthogonality does not arise if single functions are used instead of the sum of a series of such functions. However, a single function will provide only approximate solutions for the critical edge load and for the deflection under transverse load; it will provide rather poor approximations to the stresses due to transverse load. The critical load will be overestimated and the deflection under transverse load will be underestimated.

(ii) Method of M. Levy

If two opposite edges are simply supported (say $y = 0, b$) then the solution of the differential equations of the problem may be written in the form

$$w = \sum \sum a_{mn} \phi_m(x) \sin \frac{n\pi y}{b}. \tag{9.10}$$

This satisfies the boundary conditions at $y = 0, b$. The problem reduces to the solution of an ordinary differential equation for $\phi_m(x)$, which must also satisfy the boundary conditions at the edges $x = 0, a$. The function $\phi_m(x)$ is generally the sum of hyperbolic and trigonometric functions.

Functions of a form similar to equation (9.10) are also needed to define the other two variables in the problem (u and v, γ_{zx} and γ_{yz}, etc.—Table 9.1).

Procedures of this kind have often been adopted for panels with two opposite sides simply supported and the other two clamped. See, for example, Seide,[2.3, 2.9] Yen, Salerno and Hoff.[2.6]

The method is not convenient for a panel with all four edges clamped because it is not then possible to choose a simple function (comparable with $\sin(n\pi y/b)$ in equation (9.10)). which will satisfy the boundary conditions on two opposite edges.

(iii) Lagrangian Multiplier Method

In this method, deflection functions are chosen which satisfy the differential equations but not the boundary conditions (the reverse of the Rayleigh–Ritz procedure). Although individual modes do not satisfy the boundary conditions, the sum of all the separate modes may be made to do so approximately. The potential energy of the system is therefore minimized with respect to the amplitudes of the deflection functions, but the amplitudes are not independent variables. Instead they are interrelated

THE THEORY OF SANDWICH PANELS 211

by the need to satisfy the boundary conditions. This problem can be solved by the method of Lagrangian multipliers and an example can be seen in the work of Thurston[4.5] (1957). Thurston determines the buckling load N_x of an isotropic panel with all edges clamped by using the notation of Hoff[2.5] (Table 9.1) and by expressing w as the sum of a cosine series and u and v as the sums of mixed sine/cosine series. Although complicated, the method permits the evaluation of upper and lower bounds to the critical load.

9.5. Shear and Other Edge-wise Loads

Consider the effect of a uniform edge-wise shear force N_{xy} on the analysis of Chapter 6. The potential energy associated with N_{xy} is

$$V_1 = \int_A N_{xy} \frac{\partial w}{\partial x} \frac{\partial w}{\partial y} dA. \tag{9.11}$$

This term must be added to equation (6.15). In the case of a simply-supported panel it is usual to employ sinusoidal deflection functions in equation (6.17) because they satisfy the boundary conditions. Unfortunately, as substitution for w in equation (9.11) will show, the cross-product terms formed in the multiplication of the series for $\partial w/\partial x$ and for $\partial w/\partial y$ do not vanish. It is no longer possible to conduct the subsequent analysis in terms of the (m, n)th mode alone because all the modes interact with each other, and the simplicity of the method is lost.

This may be expressed differently in that buckling due to N_{xy} occurs as a mixture of all the different modes (m, n), whereas buckling due to N_x or N_y occurs in one particular mode only. An approximation to the critical value of N_{xy} may be obtained by performing an analysis analogous with Chapter 6, but including the effect of the cross-product terms, and taking into account a definite number of terms of the series (6.17). Something of this kind was done by Kuenzi, Ericksen and Zahn[1.4] (1947, 1962), but the solution is not simple. Another solution is by Seide[2.4] (1949).

In Section 7.7 also it was mentioned that single sinusoidal deflection functions fail to satisfy the differential equation (7.31) if N_{xy} is not zero.

The same difficulty arises when N_x or N_y vary in magnitude along the edges; this happens, for example, when the panel is subjected to bending in its own plane. A number of papers deal with this problem and with the problem of buckling under various combinations of edge loads applied simultaneously. See, for example, Harris and Auelmann[4.6] (1960).

9.6. Large Deflections

It is generally recognized that the ordinary theory of bending can be applied to a homogeneous plate only when the transverse displacements are small in comparison with the thickness of the plate (Timoshenko,[35.3] § 13). Otherwise, stretching of the middle plane occurs, the distribution of the forces N_x, N_y, N_{xy} throughout the plate is altered, and the more complicated large deflection theory must be used.

For a sandwich plate the restriction is more severe, and this may be demonstrated by considering two extreme cases. When the core is rigid in shear, the sandwich plate is subject to the same arguments as those applied to a homogeneous plate (except for the difference in flexural rigidity). Thus the deflections must be small compared with the thickness of the sandwich, say d. When the core is weak in shear, the faces act as two independent plates and their deflections must be small compared with the thickness of each face, t. It has already been demonstrated in Chapters 4, 5 and 6 that the parameter λ represents the transition from one extreme to the other, varying from $(-t/c)$ when the core is weak to $(+1)$ when the core is rigid in shear. Thus a simple empirical formula, (9.12), can be used to define a nominal sandwich thickness which varies from t ($G_c = 0$) to d ($G_c = \infty$).

$$t_{\text{nom}} = t\left(1+\frac{c}{d}\right)+\frac{c^2}{d}\lambda. \qquad (9.12)$$

The analyses in this book, and, indeed, in the overwhelming majority of the literature on sandwich panels, are valid only for small deflections in which the transverse displacements are less than t_{nom}. The value of λ is given for a simple isotropic sandwich by equation (6.35).

A study of large-deflection theory for sandwich plates with very thin isotropic faces and antiplane orthotropic cores is made by Alwan[5.17, 5.18] (1963, 1964) and Reissner's[4.1] early paper is formulated in terms of large deflections.

9.7. Initial Deformations

Suppose that a simply-supported panel is not perfectly flat in the unloaded state. The initial deformation w_0 can be expressed in the form

$$w_0 = \sum_{m=1}^{\infty} \sum_{n=1}^{\infty} \bar{a}_{mn} \sin\frac{m\pi x}{a} \sin\frac{n\pi y}{b}, \qquad (9.13)$$

where the amplitudes $\bar{a}_{mn}$ can, in theory, be determined from measurements of the deformation.

When loads are applied to the plate the final deformation w can be expressed in a similar form:

$$w = \sum_{m=1}^{\infty} \sum_{n=1}^{\infty} a_{mn} \sin\frac{m\pi x}{a} \sin\frac{n\pi y}{b}. \qquad (9.14)$$

Here, however, the amplitudes a_{mn} are unknown and must be found in terms of the applied load and the initial values $\bar{a}_{mn}$. This can easily be done, for example, by substitution in a modified form of the differential equation (9.1). The modification consists of replacing w on the left-hand side by $(w-w_0)$. This is because the bending and twisting moments in the plate are related to the *change* in deflection, not to the total deflection w. The modification is *not* made on the right-hand side of the equation because the edge forces N have an eccentricity of w with respect to the deflected middle plane of the plate, not $(w-w_0)$.

If the substitution is carried out for, say, $N_x \neq 0$, $q = N_y = N_{xy} = 0$, it is easy to show that:

$$a_{mn} = \frac{\bar{a}_{mn}}{1 - P/P_{cr}}, \qquad (9.15)$$

where $P = -N_x$ and P_{cr} is the critical value of P in the (m, n)th mode (equation (5.28)).

This simple relationship makes it easy to calculate the deflections and stresses in the plate once the initial irregularities are known. In particular, if the plate is approximately square and isotropic the lowest critical value of P corresponds to the mode $m=n=1$. The form of equation (9.15) shows that as P approaches P_{cr}, the amplitude a_{11} becomes large; the contributions of the other modes may therefore be neglected and only the first terms in equations (9.13) and (9.14) need to be considered. It is now possible to determine the value of the edge load, P, at which the maximum stress in the faces reaches some arbitrary value, such as the proof stress or the yield stress; this edge load may be interpreted as the maximum load which the plate will carry.

Unfortunately there are serious difficulties in the way of this interpretation. First, it is difficult to arrive at a rational assessment of the initial amplitude a_{11}, which must in any case be based on the results of extensive observations of real panels. Second, the deformations are likely to become "large" in the sense of Section 9.6 before the maximum load-carrying capacity of the panel is reached; if this happens equation (9.15) and the small-deflection theory on which it is based are no longer valid. Third, many panels are able to sustain loads substantially larger than those at which the yield or proof stress is first reached; plastic behaviour of the panel must therefore be taken into account.† As a result, a realistic account of the collapse behaviour of plates with initial irregularities should take into account large deflections and plastic behaviour.

The ordinary buckling theory for flat plates avoids these objections because it purports to do no more than predict the load at which the panel first begins to deviate from the flat condition. Deflections are always small, therefore, and the effects of plasticity may be allowed for approximately by the use of the tangent modulus for the face material.

There is no simple criterion of whether or not a plate is "flat". The load-deflection curves of two plates, one nearly flat, one less flat, are shown diagrammatically in Fig. 9.4 together with the theoretical curves for a perfectly flat plate according to the small-

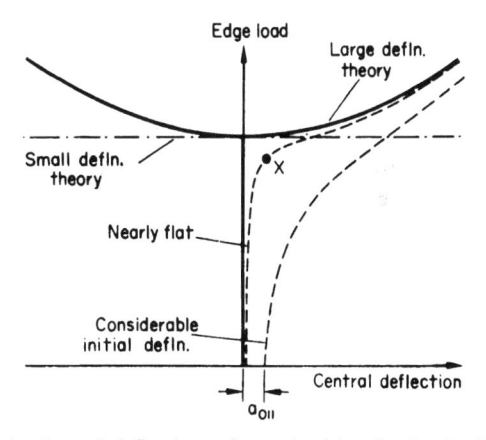

FIG. 9.4. Central deflection of panel with edge load. (From fig. 2.10, *Introduction to Structural Stability Theory*, by G. Gerard. Copyright 1962, McGraw-Hill. Used by permission of McGraw-Hill Book Company.)

---

† In *struts* the second difficulty no longer applies and the third difficulty is less important.

and large-deflection theories. The critical load P_{cr} gives a good approximation to the kink in the curve for the nearly-flat plate (marked X). The curve for the plate which has a substantial initial deformation shows no such knee; knowledge of the value of P_{cr} is not particularly useful and the effects of large-deflection and plastic behaviour need to be considered in detail.

The effects of initial deformation are examined by Chang et al.[4.7, 5.11] (1958, 1960).

CHAPTER 10

FORMULAE FOR ANALYSIS

10.1. Assumptions and Approximations

Meaningful use of the formulae in this chapter (and indeed in the rest of the book) depends on a clear understanding of the degrees of approximation which can be used to describe the behaviour of different sandwiches. This section is a summary of the assumptions and approximations which are legitimate in various circumstances. It will sometimes be convenient to reproduce equations derived elsewhere in the book; these equations are given new numbers, but the old numbers are given in square brackets so that the source may be traced readily.

Flexural rigidity

The flexural rigidity of a narrow sandwich beam is

$$D = \frac{E_f b t^3}{6} + \frac{E_f b t d^2}{2} + \frac{E_c b c^3}{12}, \qquad (10.1)\ [2.2]$$

where E_f and E_c are the face and core moduli along the axis of the beam and the dimensions are defined in Fig. 2.1.

On the right-hand side the first term may be neglected[†] in comparison with the second if:

$$\frac{d}{t} > 5{\cdot}77. \qquad (10.2)\ [2.4]$$

[†] A term p is neglected in comparison with a term q if $p < q/100$, unless otherwise stated.

If this condition is fulfilled the local bending stiffness of the faces (bending about their own separate centroidal axes) makes a negligible contribution of the flexural rigidity of the sandwich.

The third term may be neglected[†] in comparison with the second if

$$\frac{E_f}{E_c} \frac{t}{c} \left(\frac{d}{c}\right)^2 > 16\cdot 7. \qquad (10.3)\ [2.5]$$

If this condition is fulfilled the bending stiffness of the core is negligible.

Shear stress in core

The shear stress (or strain) in the core is uniform over the depth of the core if

$$\frac{E_f}{E_c} \frac{t}{c} \left(\frac{d}{c}\right) > 25. \qquad (10.4)\ [2.11]$$

For practical purposes this is the same condition as (10.3). If it is fulfilled the core is said to be of the "antiplane" type.

Shear stiffness

The shear stiffness of a sandwich beam is

$$AG = \frac{Gbd^2}{c}. \qquad (10.5)\ [2.15b]$$

If $d/c \doteqdot 1$ this may be written in the simpler form:

$$AG = Gbd. \qquad (10.6)$$

This step is valid[†] provided

$$\frac{d}{t} > 100. \qquad (10.7)$$

[†] A term p is neglected in comparison with a term q if $p < q/100$, unless otherwise stated.

FORMULAE FOR ANALYSIS

The approximation (10.6) is equivalent to neglecting the thickness of the faces when considering the geometry of the deformation of the sandwich.

Definition of a thin face

Conditions (10.2) and (10.7) may be used to define the terms used elsewhere in this book.

"Very thin" face: $\dfrac{d}{t} > 100$; $A = bd$:   neglect I_f.

"Thin" face: $100 > \dfrac{d}{t} > 5{\cdot}77$; $A = \dfrac{bd^2}{c}$:   neglect I_f.

"Thick" face: $\dfrac{d}{t} < 5{\cdot}77$; $A = \dfrac{bd^2}{c}$:   include I_f.

Interaction of thick faces and a flexible core

Where the faces are thick the shear deformation of the core is modified by the resistance of the faces to bending about their own centroidal axes (compare Figs. 2.5c and d). The effect is manifested in the factor $(1 - I_f/I)$ and in the coefficients $\psi_1 - \psi_6$ in equations (2.38) and (2.46) and Figs. 2.10 and 2.13. The error in the shear deflection caused by equating ψ_1 or ψ_4 (but not $(1 - I_f/I)$) to unity is less than 10% if ψ_1 or ψ_4 are greater than 0·90. The diagrams show that this is certainly true if

$$\theta > 15. \qquad (10.8)$$

The quantity θ depends on the span, the proportions of the sandwich and the relative stiffnesses of the face and core materials. It is defined by equation (2.39) but, for sandwiches with equal faces, Fig. 2.11 may be used to evaluate it.

Other assumptions

Shear deformations in the faces themselves are invariably neglected.

Unless otherwise stated (as in the sections on wrinkling) the core strains in the direction perpendicular to the faces are neglected.

It is generally assumed that the faces are of equal thickness and similar materials, but modifications for faces with unequal thicknesses are given where they can be made without great difficulty (see Section 2.9).

10.2. Sandwich Beams

In the formulae which follow it is assumed that the beam is narrow and that it bends anticlastically. If the beam is wide and it bends cylindrically an approximate correction may be made by replacing E_f by $E_f/(1-v_f^2)$ and D_1 by D_2.

It is also assumed that *either* the load is symmetrical *or* some restraint is applied at one end or both to prevent one face from moving bodily with respect to the other, as Fig. 2.8. Otherwise Sections 2.4 and the closing paragraphs of Section 2.5 should be consulted.

If condition (10.4) is not satisfied the shear stress is not uniform across the depth of the core and Section 2.10 is applicable.

Beams with thin or very thin faces (Section 2.3)

The direct stresses in the faces and the shear stress in the core may be calculated by the ordinary theory of bending. The total deflection is the sum of the ordinary bending deflection w_1, calculated in the ordinary way, and an additional deflection w_2 associated with shear deformation of the core. The shear deflection is found by the integration of equation (2.15a). If the beam is simply supported the shear deflection curve is equal to the bending moment diagram divided by AG (equation (2.16)). The results in two special cases, taken from Section 2.3, are as follows:

Beam with central point load W

$$\Delta = \frac{WL^3}{48D_1} + \frac{WL}{4AG}. \tag{10.9}$$

Beam with uniformly-distributed load q

$$\Delta = \frac{5qL^4}{384D_1} + \frac{qL^2}{8AG}, \qquad (10.10)$$

where $D_1 = E_f btd^2/2$, $A = bd^2/c$, Δ = central deflection.

Beams with thick faces (Sections 2.5–2.8)

The analysis of Section 2.5 must be used when the faces are thick. The method suggested involves the solution of the differential equation (2.27a) for the shear force component Q_1 associated with bending in the absence of core shear strains. The stresses in the faces and core and the additional deflections associated with core shear strains are computed indirectly from Q_1.

For a beam with a central point load W the maximum deflection, the maximum direct stress in the faces and the maximum shear stress in the core are given by equations (2.37). The equations contain the coefficients ψ_1, ψ_2, ψ_3 which depend on the overhang at the ends of the beam and on the value of θ, which is in turn a function of the span, the proportions of the sandwich and the relative stiffness of the face and core materials. ψ_1, ψ_2, ψ_3, are defined by equations (2.38) and (2.39). For large and small overhangs Fig. 2.10 presents ψ_1, ψ_2, ψ_3, in terms of θ; for sandwiches with equal faces θ can be found from Fig. 2.11.

In a similar manner equations (2.45) determine the maximum deflections and stresses in a beam with uniformly-distributed load in terms of coefficients ψ_4, ψ_5, ψ_6 (equations (2.46)) which may be obtained from Fig. 2.13.

10.3. Sandwich Struts

In the formulae which follow it is assumed that the strut is narrow and that it bends anticlastically. If the strut is wide and its bends cylindrically an approximate correction may be made by replacing E_f by $E_f/(1-\nu_f^2)$ and D_1 by D_2.

In any analysis of the buckling of a sandwich strut the buckling stress should be taken as the lower of (i) the stress given by the formulae in this section and (ii) the wrinkling stress given in Section 10.4. In fact the overall instability and wrinkling instability interact as described in Section 8.10, but the interaction is of little practical importance.

Consideration should also be given to the possibility of local instability of the elements of a corrugated-core sandwich or the intracellular buckling of the faces of a honeycomb-core sandwich (Section 8.9).

Struts with thin or very thin faces (Section 3.1)

The critical load of a pin-ended strut is

$$P = \frac{P_E}{1 + P_E/AG}, \qquad (10.11)\ [3.6]$$

where $P_E = \pi^2 D_1/L^2$, $D_1 = E_f btd^2/2$, $A = bd^2/c$.

Equation (10.11) is represented by the curve HKC in Fig. 3.2.

The critical load of a strut with other boundary conditions is also given by equation (10.11) provided P_E is replaced by the critical load for the case $G = \infty$. For example, when both ends are fixed P_E should be replaced by $4P_E$.

Struts with thick faces (Sections 3.2 and 3.3)

In general the critical load is obtained from consideration of the solution of the differential equation (3.8) (as in Section 3.3).

In particular the critical load of a pin-ended strut with the end-conditions illustrated in Fig. 3.3 is

$$P = P_E \frac{1 + (P_{Ef}/P_c) - (P_{Ef}/P_c) \cdot (P_{Ef}/P_E)}{1 + (P_E/P_c) - (P_{Ef}/P_c)}, \qquad (10.12)\ [3.13]$$

where $P_E = \pi^2 D_1/L^2$, $P_{Ef} = \pi^2 E_f I_f/L^2$, $P_c = AG = Gbd^2/c$, $D_1 = E_f btd^2/2$.

Equation (10.12) is represented by the curve ABC in Fig. 3.2.

Equations (3.11) and (3.14) are equivalent to (10.12) and may be more convenient to use in certain problems.

10.4. Wrinkling

The wrinkling formulae which follow refer to the possibility of short-wavelength instability of the faces of sandwich members in the manner illustrated in Fig. 8.4. When the panels are wide E_f should be replaced by $E_f/(1-v_f^2)$.

Pure wrinkling of axially loaded struts and panels (isotropic core) (Section 8.3)

The stress in the faces at which wrinkling occurs is

$$\sigma = B_1 E_f^{\frac{1}{3}} E_c^{\frac{2}{3}}. \qquad (10.13) \ [8.12a]$$

The form which the wrinkling takes is determined by the Poisson's ratio of the core (v_c) and the non-dimensional quantity:

$$\varrho = \frac{t}{c} \left\{ \frac{E_f}{E_c} \right\}^{\frac{1}{3}}. \qquad (10.14)$$

There exists a limiting value if ϱ, defined by equation (10.15). When ϱ is less than this value, antisymmetric wrinkling is possible (Fig. 8.4, II) and B_1 may be found from Fig. 8.6. When ϱ is greater than the critical value, symmetric wrinkling is possible (Fig. 8.4, III) and B_1 may be found from Fig. 8.7.

$$\varrho_1 = \frac{1}{2} \left\{ \frac{1-v_c}{1+v_c} \right\}^{\frac{1}{3}}. \qquad (10.15)$$

Equation (10.13) is sometimes written in an alternative form:

$$\sigma_1 = B_1'(E_f E_c G_c)^{\frac{1}{3}} \qquad (10.16)$$

where $B_1' = B_1[2(1+v_c)]^{\frac{1}{3}}$.

When $\varrho < 0.2$ the value of B_1 is the same for symmetric and antisymmetric wrinkling and it is independent of ϱ; this value may be obtained from Table 10.1. The table also shows the absolute minimum value of B_1, which occurs in antisymmetric wrinkling when $\varrho = \varrho_1$.

TABLE 10.1

v_c	$\varrho < 0.2$		Absolute minimum values	
	B_1	B_1'	B_1	B_1'
0	0·63	0·80	0·50	0·63
0·25	0·58	0·79	0·47	0·64
0·5	0·54	0·78	0·47	0·68

Pure wrinkling of the compression face of a sandwich beam (isotropic core) (Section 8.3)

The situation corresponds approximately with that in Fig. 8.4 (I), with the rigid base representing the tensile face of the sandwich. The wrinkling stress is again given by equation (10.13), but the value of B_1 should be taken instead from Fig. 8.5.

Tensile failure of the bond between the faces and the core (isotropic core) (Section 8.7)

When a face has initial irregularities, the application of compressive stress in the plane of the faces causes the amplitude of the irregularities to be increased. At the same time a tensile stress develops in places between the face and the core. When this tensile stress reaches its ultimate value, σ_u, the bond between the face and the core fails. It is shown in Section 8.7 that the minimum compressive stress in the face at which this can happen is given by

$$\sigma = B_2 E_f^{\frac{1}{3}} E_c^{\frac{2}{3}}. \qquad (10.17) \ [8.34]$$

The value of B_2 is always less than that of B_1 in equation (10.13). It is a function of the quantity ϱ (equation (10.14)) and a non-dimensional coefficient k (defined by equation (8.33)) which depends on the amplitude of the initial irregularities, the ultimate tensile strength of the interface and certain assumptions discussed in Section 8.7. The value of B_2 may be obtained from Fig. 8.11 if k is known. Unfortunately k cannot easily be calculated; instead it is necessary to perform compression tests on short sandwich columns in order to plot experimental points in Fig. 8.11 and hence to determine which value of k gives the best fit. Further details are given in Section 8.7.

*Wrinkling and tensile bond failure of axially loaded struts
  and panels (orthotropic core)* (Section 8.8)

As with isotropic cores, antisymmetric wrinkling (Fig. 8.4, II) is more likely to occur than symmetric wrinkling (Fig. 8.4, III). An exception to this rule is possible when the core transverse stiffness E_z is rather less in relation to G_{zx} than it would be for an isotropic core. Such a core is unlikely to be used in practice.

The report by Norris[6.3] includes a diagram analogous with Fig. 8.11 from which the antisymmetric wrinkling stress may be obtained when the core is orthotropic. The diagram is valid for plane stress or plane strain conditions and it also permits the estimation in terms of k of the face stress at which tensile bond failure occurs. A similar diagram appears in ref. (30.2) (p. 51).

A special case of an orthotropic core is the antiplane core ($E_x = E_y = 0$) in which the transverse stiffness E_z is infinite. It is shown in Section 8.6 that wrinkling cannot occur in a sandwich with a core of this type.

Local instability

In a sandwich with a honeycomb core there is a possibility of the faces buckling into the cells of the core (intracellular buckling). Equations (8.35a and b) have been proposed for the prediction of such instability.

The local instability of plate elements of a sandwich with a corrugated core is beyond the scope of this book, but design curves are available in ref. (5.12) (part III, chapter 5).

It is strongly recommended that any formulae used for the prediction of local instability should be verified by tests on short column specimens.

10.5. Buckling of Simply-supported Panels with Edge Forces in the x-direction (Sections 5.4, 6.7 and 7.8)

The notation is defined in Figs. 5.1, 5.2 and 6.1.

Isotropic thin faces; isotropic core

The critical value of the edge force per unit length in the x-direction is

$$P = \frac{\pi^2 D_2}{b^2} K_1, \qquad (10.18)\ [5.29a]$$

where b is the width of the panel in the y-direction, D_2 is the flexural rigidity $E_f t d^2/2(1-\nu_f^2)$ and K_1 is a dimensionless coefficient which can be obtained from Fig. 5.4 in terms of a/b (the length/width ratio of the panel), and a quantity ϱ.

ϱ here represents the ratio of the flexural and shear rigidities and it is defined as follows:

$$\varrho = \frac{\pi^2}{2(1-\nu_f^2)} \frac{E}{G} \frac{tc}{b^2}. \qquad (10.19)\ [6.34b]$$

(It is important to note that ϱ is taken from equation (6.34b), not equation (5.25c); the latter is valid only for very thin faces but the former can be used for thin or very thin faces without extra complication. Figure 5.4 is valid for either ϱ.)

FORMULAE FOR ANALYSIS

Isotropic thick faces; isotropic core

The critical value of the edge force per unit length in the *x*-direction is

$$P = \frac{\pi^2 D_2}{b^2} K_2. \qquad (10.20) \quad [6.36a]$$

b and D_2 are defined as for equation (10.18), but K_2 must be obtained from Fig. 6.3 in terms of a/b and ϱ (as above) and also t/d.

It is clear from the diagram that K_2 is affected very little by t/d except in two extreme cases. In considering these cases it is convenient to define a point such as X in Fig. 6.3 by the coordinates $K_2 = \overline{K}_2$, $a/b = \bar{r}$. The point X represents the minimum of the curve $m = 1$. The value of $\bar{r}$ may be found from Fig. 10.1 in terms of ϱ and t/d.

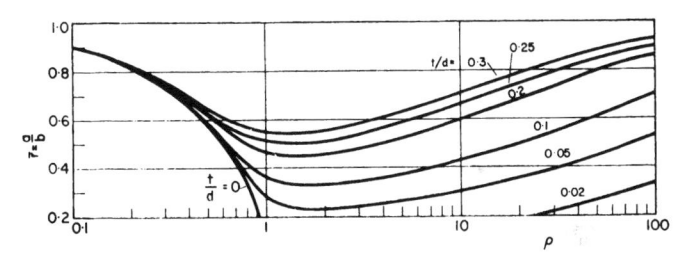

FIG. 10.1. Values of a/b at which the buckling coefficient K_2 is minimum when $m = 1$ (e.g. point X in Fig. 6.3).

Case (i)

When the panel is very short (i. e. $a/b < \bar{r}$) K_2 may be obtained from equation (6.36c) by inserting the desired values of a/b, t/d and ϱ, also $m = n = 1$. This is equivalent to the more accurate determination of a point on the curve to the left of X in Fig. 6.3.

Case (ii)

When the panel is longer (i.e. $a/b > \bar{r}$) and ϱ is large (greater than 2, say), Fig. 6.3 implies that K_2 is approximately equal to

$\bar{K}_2$ and is independent of a/b. In other words, the curve to the right of X is nearly horizontal. This K_2 may be obtained from equation (6.36c) by inserting the desired values of t/d and ϱ, also $a/b = \bar{r}$, $m = n = 1$.

A convenient procedure in this case is to determine the critical load from equation (10.21) (an alternative form of (10.20)).

$$P = \left\{ \frac{\pi^2}{2(1-v_f^2)} \frac{E_f d^2 t}{b^2} \frac{(\bar{r}+\bar{r}^{-1})^2}{1+\varrho(1+\bar{r}^{-2})} \right\} + \left\{ \frac{\pi^2}{6(1-v_f^2)} \frac{E_f t^3}{b^2} (\bar{r}+\bar{r}^{-1})^2 \right\}. \tag{10.21}$$

Orthotropic panels with thin faces

The critical value of the edge force per unit length in the x-direction is

$$P = \frac{\pi^2}{b^2} \frac{D_x}{(1-v_x v_y)} K_3, \tag{10.22} \quad [7.57a]$$

where b is the width of the panel in the y-direction, D_x is the *anticlastic* flexural rigidity in the x-direction, v_x and v_y are the Poisson's ratios defined in terms of curvatures of the plate.

The flexural rigidities D_x, D_y, the torsional rigidity D_{xy}, the shear stiffnesses D_{Qx} and D_{Qy} and the Poisson's ratios v_x, v_y are defined in equations (7.1)–(7.6). The actual evaluation of these quantities is explained in Section 7.6

In general it is necessary to obtain the coefficient K_3 from equation (7.57b), with $n = 1$, $P_y = 0$ and with the integer m chosen to make K_3 minimum. In some common special cases, however, K_3 may be obtained directly from graphs as follows.

Isotropic thin faces; orthotropic (honeycomb) cores

K_3 is given in Fig. 7.4 in terms of a/b (the ratio of the length of the plate to the width) and $D_x/b^2 D_{Qx}$, which represents the ratio of the flexural and shear stiffnesses in the x-direction in non-dimensional form.

The three graphs are for the cases $D_{Qx}/D_{Qy} = 0.4$, 1.0 and 2.5 and they cover the range of values to be expected in most honeycomb cores. It should be remembered that conventional honeycomb cores have greater shear stiffness in the L-direction than in the W-direction (Fig. 12.3); whether D_{Qx}/D_{Qy} is greater or less than unity depends on the orientation of the core with respect to the direction of the edge load.

Isotropic thin faces; orthotropic (corrugated) cores

When the corrugations run in the x-direction (the direction of the edge load) the value of K_3 can be found from Fig. 7.5 according to the ratios a/b and $D_y/b^2 D_{Qy}$. The shear stiffness in the x-direction is assumed to be infinite. Graphs are plotted for the four ratios $D_x/D_y = 1\cdot 0, 1\cdot 25, 1\cdot 5, 1\cdot 75$; D_y is essentially the flexural rigidity associated with the faces alone, but D_x includes the flexural rigidity of the corrugated core. The ratio D_x/D_y evidently depends on the geometry of the core and the relative thicknesses of the faces and core, as described in Section 7.6.

Figure 7.6 permits the evaluation of K_3 when the corrugations run across the direction of loading; it is directly comparable with Fig. 7.5a and, as might be expected, it provides lower values of K_3.

10.6. Bending of Simply-supported Panels with Uniform Transverse Load (Sections 5.4, 6.7 and 7.9)

The results which are quoted in this section and in the next are valid only if the transverse displacements of the panel are small. The significance of this limitation is discussed in Section 9.6.

The various coefficients α, β, which appear in equations (10.23), (10.24) and (10.25) are plotted in graphs elsewhere in the book. The plotted points have been obtained by evaluating the series for α and β up to and including the term $m = n = 23$. This appears to give adequate accuracy for practical purposes, at least for the circumstances covered by the graphs.

Some of the values of β in Fig. 5.5 may be compared with results obtained by Timoshenko[35.3] (chapter 5, table 5) by a more sophisticated method. Timoshenko's results are for a homogeneous plate but the bending moments and shear forces in such a plate are identical with those in an isotropic sandwich plate; the deflections are also comparable when ϱ is zero.

Isotropic thin faces; isotropic core; panel of any length and width

The maximum transverse displacement due to a uniform transverse load q per unit area is

$$w_{\max} = \frac{qb^4}{D_2}(\beta_1 + \varrho\beta_2), \qquad (10.23)\;[5.35a]$$

where b is the width of the panel in the y-direction, D_2 is the flexural rigidity $E_f t d^2/2(1-v_f^2)$ and ϱ is the ratio of the flexural rigidities according to equation (10.19) [6.34b]. As in the buckling formula, ϱ is taken from equation (6.34b) rather than from equation (5.25c) so that equation (10.23) may be used for sandwiches with thin faces as well as for those with very thin faces.

The coefficients β_1 and β_2 given in Fig. 5.5 for panels with different ratios a/b.

In a similar manner, the membrane stresses in the faces and the shear stresses in the core may be found from equations (5.39) in terms of the coefficients β_3–β_7 which are also plotted in Fig. 5.5.

Isotropic thick faces; isotropic core; square panel

The maximum transverse displacement due to a uniform transverse load q per unit area is

$$w_{\max} = \frac{qb^4}{D_2}\alpha_1, \qquad (10.24)\;[6.38a]$$

where b is the width of the panel in the y-direction and D_2 is equal to $E_f t d^2/2(1-v_f^2)$.

The coefficient α_1 is plotted in Fig. 6.4 in terms of ϱ (equation (10.19)) and t/d.

In a similar manner, the membrane stresses in the faces, the local bending and twisting stresses in the faces and the shear stresses in the core may be found from equations (6.38) in terms of the coefficients α_3–α_{10} which are also plotted in Fig. 6.4.

Orthotropic square panels with thin faces

The maximum transverse displacement due to a uniform transverse load q per unit area is

$$w_{max} = \frac{qb^4}{D_x} \beta_1, \qquad (10.25)\ [7.59]$$

where b is the width of the panel in the y-direction, D_x is the *anticlastic* flexural rigidity in the x-direction and the coefficient β_1 is defined by the series (7.65a).

For panels with isotropic faces and orthotropic (honeycomb) cores β_1 may be obtained from Fig. 7.7.

For panels with isotropic faces and orthotropic (corrugated) cores β_1 may be obtained from Fig. 7.8.

The bending and twisting moments M_x, M_y, M_{xy} and the shear forces Q_x, Q_y are defined by equations (7.60)–(7.64) in terms of coefficients β_2–β_6 which are also plotted in Figs. 7.7 and 7.8.

Panels which are not square

The coefficients α and β in equations (10.24) and (10.25) have been plotted for square plates only. They may be evaluated for panels of any shape but it is not practicable to plot so many values here. If the ratio a/b is greater than about 3, however, it is permissible to consider the central strip of the panel, of width dx, as a simply-supported beam of span b. The methods of Chapter 2 and Section 10.2 may then be applied to determine the deflection, the face stress σ_y and the core stress τ_{yz}. This procedure may not be valid if the panel is much stiffer in the x-direction than it is in the y-direction.

10.7. Bending of Simply-supported Panels with Uniform Transverse Load Combined with Edge Forces in the x-direction

Theoretically the deflections and stresses may be obtained by the methods of the previous section provided each (m, n)th term in the series in equations (5.35), (5.40), (7.65) and Table 6.1 is multiplied by the factor

$$\frac{1}{1-(P/P_{mn})},$$

where P is the load per unit length applied in the x-direction and P_{mn} is the critical value of P associated with buckling in the m, nth mode. The axial stress in the faces due to P must be added to the result.

In practice the temptation is to take only the first term of each series, in which case a solution can be obtained fairly easily. Such a procedure is not too inaccurate when used for deflections provided a/b is not much larger than unity and P is only a small proportion of P_{11}. The results for the membrane stresses in the faces are only crude approximations and those for the other stresses are even less reliable.

10.8. Modifications for Faces of Unequal Thickness or Dissimilar Materials

The principal beam and strut equations are unchanged provided (EI) and (EI_f) are interpreted in the following way:

$$(EI) = \frac{bd^2 E_1 E_2 t_1 t_2}{E_1 t_1 + E_2 t_2} + \frac{b}{12}(E_1 t_1^3 + E_2 t_2^3), \qquad (10.26a)$$

$$(EI_f) = \frac{b}{12}(E_1 t_1^3 + E_2 t_2^3), \qquad (10.26b)$$

where the suffixes 1 and 2 refer to the upper and lower faces respectively. The shear stiffness remains unchanged at Gbd^2/c, where d represents as usual the distance between the centroids of the upper and lower faces. Equations (10.26) represent a slight generalization of the results in Section 2.9.

The isotropic plate buckling and bending equations (10.18) and (10.23) are unchanged provided the upper and lower faces have the same Poisson's ratio v_f and the following alterations are made:

$$D_2 = \frac{E_1 E_2 t_1 t_2 d^2}{(1-v_f^2)(E_1 t_1 + E_2 t_2)}, \tag{10.27a}$$

$$\varrho = \frac{\pi^2}{b^2(1-v_f^2)} \frac{E_1 E_2 t_1 t_2 c}{G(E_1 t_1 + E_2 t_2)}. \tag{10.27b}$$

The orthotropic plate buckling and bending equations (10.22) and (10.25) are unchanged provided the upper and lower faces have the same Poisson's ratios and D_x, D_y, D_{xy}, D_{Qx}, D_{Qy} are interpreted as is Section 7.6 (iii).

The modifications outlined in this section are strictly valid only in the particular circumstances set out in Section 9.2 (see especially equation (9.4)). Small departures from these special circumstances should not have serious practical consequences.

10.9. Buckling and Bending of Panels with Other Types of Load or Boundary Conditions

The provision of comprehensive design data for all load cases and boundary conditions is beyond the scope of this book; however, the publications listed in Table 10.2 cover most of the important cases which are likely to arise in practice. The table indicates the kind of information which may be obtained from each source. Some publications cover a wide variety of situations and the table must therefore be treated as a rough guide rather than as a rigorous classification.

TABLE 10.2. CIRCLES INDICATE THAT INFORMATION *Is* PROVIDED

References	Faces						Core			Edges			Loads					Results		
	Isotropic	Orthotropic	Equal thickness only	Thickness can be unequal	Similar materials only	Materials can be different	Isotropic	Orthotropic (honeycomb)	Orthotropic (corrugated)	All simply supported	Some s.s., some clamped	All clamped	Transverse (mainly u.d.)	N_x or N_y (not both)	N_{xy}	Combined edge loads	Edge and transverse loads	Buckling loads	Transverse deflections	Stresses
1.18	○	○		○		○		○	○	○				○				○		
1.17	○	○		○		○	○	○		○	○	○		○				○		
5.12	○	○		○		○	○	○	○	○	○	○	○	○	○	○	○	○	○	○
5.8 and 5.9	○			○	○		○	○	○	○	○	○		○				○		
5.7	○			○		○		○	○	○				○				○		
5.7	○			○		○		○	○							○				○
4.6	○		○		○			○	○					○	○	○		○		
1.7	○			○		○	○				○	○		○				○		
1.8	○			○	○		○			○			○						○	
1.9	○			○	○		○			○			○						○	○
1.13	○			○	○			○		○			○						○	○
1.15	○			○	○		○	○		○	○		○					○	○	
1.16		○		○	○		○	○		○			○					○	○	
1.4																	○	○	○	
2.3	○		○		○		○				○			○				○	○	
2.2	○		○		○		○			○				○				○	○	
2.9	○		○		○				○	○	○			○				○	○	
2.4	○		○		○		○			○						○		○		
2.7	○		○		○		○			○			○							○

CHAPTER 11

DESIGN OF SANDWICH BEAMS, STRUTS AND PANELS

11.1 Introduction

The process of trial and error is often the most effective method of designing sandwich panels. Elaborate methods of optimum design have occasionally been proposed in which the proportions of a sandwich with specified loads and spans are adjusted with minute precision in order to save the last ounce of weight. In reality, however, the choice of faces and cores is not infinite; face materials may be available in relatively few gauges or standard thicknesses; core materials may be restricted in the choice of thickness and density. In such cases it can be expedient to use a computer (which would otherwise be required for the preparation of optimum design charts) to check the strength and stiffness of a selection of practical sandwiches and to choose the lightest which will perform the desired task. This is particularly true when effects such as plasticity are to be taken into account and it avoids the embarrassment engendered by an optimum structure with impossibly thin faces.

All the same, it is convenient to have methods of design which can indicate roughly where the process of trial and error should begin. In the nature of things, these design methods need not be as precise as the final analysis or check calculation. Advantage can be taken of this to take short cuts or to make approximations which are not acceptable in the check calculation. Indeed, unless short cuts *are* taken, the design method is likely to be more cumbersome than the process of analysis and its use will be correspondingly restricted.

One short cut is to ignore completely any effects due to the thickness of the faces. A sandwich with thick faces and a weak core is by definition an inefficient sandwich, because the faces are well on the way to working as two independent beams, struts or panels. Furthermore, the difference between "thin" and "very thin" faces as defined in Section 10.1 is merely the difference between Gd^2/c and Gd as the shear stiffness; in terms of approximate design this difference is small enough to be neglected and the faces may be treated as "very thin".

There are three main design processes. In the first and simplest the core and face materials are specified, as is the thickness of the faces. The problem is to determine the necessary core thickness. In many building structures and in semi-structural applications the loads are light and a truly optimum design (in terms of weight or cost) would lead to impractically thin faces. Consequently it is convenient to begin by choosing the thinnest face which can be used (in terms of robustness, fire-resistance, weathering, etc.) and then finding the thinnest core which can be used with it.

In the second design process the core and face materials are specified but the thicknesses of the faces and the core are to be found. Generally there is a whole range of combinations of face and core thicknesses which will provide adequate strength or stiffness and the problem is to choose the combination which provides the sandwich with the lowest weight (or cost). It is this process which is usually referred to as optimum design, or minimum-weight design. It is more likely to be used for aero structures where weight-saving is vital and where the fabricator is willing to take extra trouble to obtain non-standard sizes to fit the design.

The third design process is similar to the second except that the core density is to be chosen as well as the face and core thicknesses. It is usual to assume that the strength and stiffness of the core are directly proportional to the density. This type of problem will not be considered here.

11.2. Determination of Core Thickness

Beam with uniformly distributed load

Consider a simply-supported sandwich beam which is required to support a uniformly distributed load q per unit length at failure. The load q is equal to the product of the working load and a suitable load factor. The face thickness t and the face and core materials are predetermined and it is desired to find d (approximately the core thickness if the faces are very thin).

The maximum bending moment is $qL^2/8$ and the maximum stress in the faces is therefore $qL^2/8bdt$. This stress must not exceed the ultimate strength of the material. Also, it must not exceed the wrinkling stress. For the present purpose the wrinkling stress may be calculated from equation (10.13) with the value $B_1 \doteq 0.55$ (Fig. 8.5). If the lesser of the ultimate strength and the wrinkling stress is denoted by σ_1, then

$$\frac{qL^2}{8bdt} \leq \sigma_1. \tag{11.1}$$

The maximum shear force is $qL/2$ and the shear stress in the core is $qL/2bd$. If the ultimate shear strength of the core is τ_1, then

$$\frac{qL}{2bd} \leq \tau_1. \tag{11.2}$$

The maximum deflection is given by equation (10.10). There may be a restriction on the deflection which can be expressed as a limiting ratio Δ_1/L at failure. Then

$$\frac{10qL^3}{384Ebtd^2} + \frac{qL}{8bdG} \leq \frac{\Delta_1}{L}. \tag{11.3}$$

The required core depth d is the least of the values which can be obtained from these three equations, viz.:

$$d \geq \frac{qL^2}{8bt\sigma_1},\qquad(11.4)$$

$$d \geq \frac{qL}{2b\tau_1},\qquad(11.5)$$

$$d \geq \frac{qL^2}{16b\,G\Delta_1}\left\{1+\sqrt{\left(1+\frac{20}{3}\frac{b}{t}\frac{G^2\Delta_1}{Eq}\right)}\right\}.\qquad(11.6)$$

Pin-ended strut

Again the face thickness t and the face core materials are predetermined and it is desired to find d to enable the strut just to support a load P over a length L. The load P is the product of the working load and a suitable load factor.

First, if there is any chance of $(t/d)(E_f/E_c)^{\frac{1}{3}}$ falling in the range indicated in Fig. 8.6, the chosen thickness t must be adequate to support P without wrinkling. For this purpose the wrinkling stress may be calculated initially from equation (10.13) with the value $B_1 = 0.5$ (Fig. 8.6). Obviously the chosen face thickness t must also be large enough to avoid the possibility of crushing or yielding of the face material.

The critical load is given by equation (10.11) and it must be greater than the specified load P. Thus:

$$\frac{(\pi^2 Ebtd^2)/2L^2}{1+(\pi^2 Etd)/2L^2G} \geq P \qquad(11.7)$$

or
$$d \geq \frac{P}{2bG}\left\{1+\sqrt{\left(1+\frac{8}{\pi^2}\frac{L^2b}{t}\frac{G^2}{EP}\right)}\right\}.\qquad(11.8)$$

Isotropic panel with uniform transverse load

Once again the face thickness t and the face and core thicknesses are predetermined. The minimum approximate core depth d is to be evaluated which will just enable the panel to support a uniform transverse pressure q. The pressure q is equal to the product of the working pressure and a suitable load factor.

As in the beam problem, the face stress at failure is limited to σ_1, which may be the ultimate strength of the face material or the wrinkling stress, whichever is the lower. The ultimate strength of the core material is τ_1 and the deflection at failure is limited to Δ_1. It is assumed that Δ_1 is equal to the limiting deflection at working load multiplied by the load factor.

It may be assumed that $a > b$ without loss of generality, in which case the face stress σ_y is greater than the face stress σ_x and the core stress τ_{yz} is greater than the core stress τ_{zx}. Hence, from equations (5.35a) and (5.39),

$$\frac{2qb^4(1-\nu_f^2)}{Etd^2}\left\{\beta_1 + \frac{\pi^2}{2(1-\nu_f^2)}\frac{E}{G}\frac{td}{b^2}\beta_2\right\} \le \Delta_1, \quad (11.9)$$

$$\frac{qb^2}{dt}(\beta_4 + \nu_f\beta_3) \le \sigma_1, \quad (11.10)$$

$$\frac{qb}{d}\beta_7 \le \tau_1. \quad (11.11)$$

The coefficients β may be obtained from Fig. 5.5 for any particular ratio a/b. The required approximate core depth d is the least of the values which can be obtained from these three relationships, viz:

$$d \ge \frac{qb^2\pi^2\beta_2}{2G\Delta_1}\left\{1 + \sqrt{\left(1 + \frac{8\beta_1 G^2 \Delta_1(1-\nu_f^2)}{Etq\pi^4\beta_2^2}\right)}\right\}, \quad (11.12)$$

$$d \ge \frac{qb^2}{\sigma_1 t}(\beta_4 + \nu_f\beta_3), \quad (11.13)$$

$$d \ge \frac{qb}{\tau_1}\beta_7. \quad (11.14)$$

If the face material is weak in shear in the xy-plane an additional condition can be obtained from equation (5.39c):

$$d \geq \frac{qb^2(1-v_f)}{\tau_{xy}t}\beta_5,\qquad(11.15)$$

where τ_{xy} is the ultimate shear strength of the faces.

Isotropic panel with uniform edge load

The face thickness t and the face and core materials are predetermined. It is desired to select the minimum approximate core depth d which will just enable the panel to support a uniform edge load P (per unit length) in the x-direction. The load P is the product of the working load and a suitable load factor.

As in the strut problem it is necessary to check first that the face thickness is adequate to support P without wrinkling or failure of the face material.

The critical load is given by equation (5.29a) and it must be greater than the specified load. Before equation (5.29a) can be used it is necessary to know the appropriate value of m to be used in equation (5.29c). Provided it is known that ϱ will be very small (i.e. the core will be quite stiff in shear) the value of m can be obtained from Fig. 5.4 for any particular value of a/b, using the curve $\varrho = 0$. For example, if $a/b < \sqrt{2}$, m is unity; if $\sqrt{2} < a/b < \sqrt{6}$, $m = 2$; if $\sqrt{6} < a/b < \sqrt{12}$, $m = 3$ and so forth. Once the value of m has been selected, equation (5.29a) may be written in the following form:

$$\frac{\dfrac{\pi^2}{b^2}\dfrac{Etd^2}{2(1-v_f^2)}\cdot\left\{\dfrac{mb}{a}+\dfrac{a}{mb}\right\}^2}{1+\left\{\dfrac{m^2b^2}{a^2}+1\right\}\dfrac{\pi^2}{2(1-v_t^2)}\dfrac{E}{G}\dfrac{td}{b^2}} \geq P.\qquad(11.16)$$

Or,

$$d \geq \frac{P}{2G\left\{1+\dfrac{a^2}{m^2b^2}\right\}}\left\{1+\sqrt{\left(1+\dfrac{8b^2G^2(1-v_f^2)}{\pi^2EtP}\dfrac{a^2}{m^2b^2}\right)}\right\}.$$

$$(11.17)$$

If ϱ is not very small it is difficult to estimate the value of m from Fig. 5.4, at least near the intersections of the curves. If ϱ is large ($\varrho > 0.3$, say) it is nearly impossible to do so.

11.3. Optimum Design: Determination of Core and Face Thickness for Minimum Weight (or Cost)

A particularly good introduction to the problem of optimum design is presented by Kuenzi[21.1] and the remarks which follow are based on his study.

Suppose that the bending stiffness D of a sandwich beam (width b) is specified, as are the materials to be used for the faces and the core. The bending stiffness is defined by equation (11.18a) and the combined weight of the faces and core (per unit area) is given by equation (11.18b), in which μ_f and μ_c are the densities of the face and core materials respectively.

$$D = \frac{Ebtd^2}{2}, \qquad (11.18a)$$

$$w = \mu_c d + 2\mu_f t. \qquad (11.18b)$$

The faces are assumed to be very thin; it is therefore in order to take d as the core thickness and to neglect the local bending stiffnesses of the faces.

The thicknesses t and d are to be adjusted to satisfy equation (11.18a) and, at the same time, to provide a minimum value for w. The weight of the adhesive is constant and so omitted from the calculation.

Elimination of t from equations (11.18a) and (11.18b) permits the expression of w in terms of d:

$$w = \mu_c d + \frac{4\mu_f D}{Ebd^2}.$$

For a minimum value of w with respect to d the following conditions must be satisfied:

$$\frac{dw}{dd} = \mu_c - \frac{8\mu_f D}{Ebd^3} = 0$$

or
$$d^3 = \frac{8\mu_f}{\mu_c} \frac{D}{bE}. \tag{11.19}$$

This is the optimum core depth. The ratio of the weight of the core to the combined weight of the faces is

$$\frac{\mu_c d}{2\mu_f t} = \frac{\mu_c d}{2\mu_f} \cdot \frac{E d^2 b}{2D} = 2. \tag{11.20}$$

This fact can be used as a quick check on the efficiency of any given sandwich; for example, a construction in which the core is only a quarter of the weight of the faces is not likely to be very efficient in terms of bending stiffness.

If the bending *strength* is specified instead of the bending *stiffness*, the face and core thicknesses must satisfy this equation:

$$M = \sigma_1 tbd. \tag{11.21}$$

Here M is the bending moment which the beam must carry at failure and σ_1 is a limiting stress such as the ultimate strength of the face material or the wrinkling stress (whichever is the lower). For the pressent σ_1 is treated as a constant. Elimination of t from equations (11.18b) and (11.21) expresses the weight of the sandwich as a function of d:

$$w = \mu_c d + \frac{2\mu_f M}{\sigma_1 bd}. \tag{11.22}$$

Minimization of w with respect to d yields the optimum core depth and shows that the weight of the core should equal the combined weight of the faces:

$$d^2 = \frac{2\mu_f M}{\sigma_1 b \mu_c}; \qquad \frac{\mu_c d}{2\mu_f t} = 1. \tag{11.23a, b}$$

If it is known or suspected that failure of the face will occur as a result of local instability (e.g. by buckling into the cells of a honeycomb of fixed cell dimensions) the limiting stress σ_1 is not constant but proportional to t^2:

$$\sigma_1 = kt^2. \qquad (11.24)$$

Equation (11.21) is now

$$M = kt^3 db. \qquad (11.25)$$

Elimination of d from equations (11.18) and (11.25) and minimization of the weight w with respect to t provides the optimum face thickness and shows that the core should weigh one-third of the combined weight of the faces:

$$t^4 = \frac{3}{2} \frac{\mu_c}{\mu_f} \frac{M}{kb}; \qquad \frac{\mu_c d}{2\mu_f t} = \frac{1}{3}. \qquad (11.26a, b)$$

The original report by Kuenzi[21.1] deals with faces of unequal thickness and dissimilar materials and also with cores in which the density is allowed to vary. Many permutations of the procedure are possible and Kuenzi uses it for an investigation of the optimum proportions of a simply-supported panel which will just support a given edge load in the x-direction. If the core is stiff in shear the critical load of a panel with a fixed a/b ratio is directly proportional to the flexural rigidity. It has already been shown that the most efficient sandwich in terms of flexural rigidity is one in which the weight of the core is twice the combined weight of the faces. This is also true of the panel with the prescribed critical load. The conclusion is less straightforward when the core is not very stiff in shear but in that case the weight of the core in the optimum sandwich is somewhat less than twice the combined weight of the faces.

A different approach is adopted by Allen.[25.3] The weight of the sandwich, as defined by equation (11.18), can be viewed as the height of an inclined plane above horizontal axes d, t (Fig. 11.1). Each of the curves A, B, C in the horizontal plane represents some relationship between d and t imposed by a requirement such as the limitation of face stress, buckling load, core shear stress, deflection, etc. The curves a, b, c in the inclined plane are vertically above A, B, C. Points such as P (on the side of a, b, c remote from the origin) represent possible sandwiches; points such as Q (on the side of a, b, c near the origin) represent sandwiches which are inadequate in some way. The object of the analysis is to select a point P as near to the origin and as far down the inclined plane as possible without crossing the curves a, b, c. Much depends on

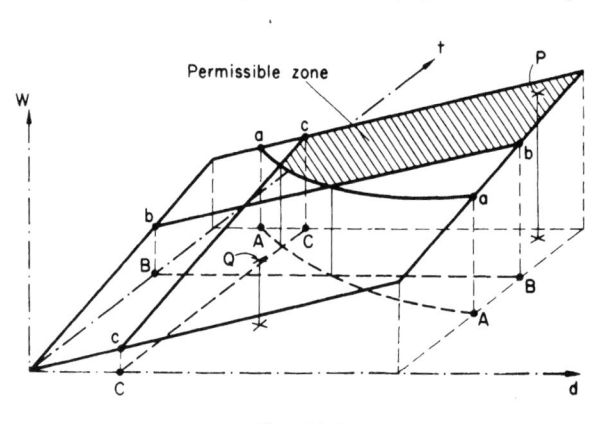

FIG. 11.1.

the way the curves intersect (if they intersect at all); this determines which combinations of physical limitations are likely to govern the design. The analysis is straightforward and it can cope with a large variety of design restrictions, but it is not really suitable for calculations by hand.

A selection of papers on other aspects of optimum design is included in the list of references under an appropriate heading. It is worth noting that any analysis which provides a minimum weight design can also be used to provide a minimum cost design if the costs of the face and core materials (per unit volume) are substituted for the densities of these materials.

CHAPTER 12

PROPERTIES OF MATERIALS USED IN SANDWICH CONSTRUCTION: METHODS OF TESTING

12.1. Face Materials

Almost any structural material which is available in the form of thin sheet may be used to form the faces of a sandwich panel. Panels for high-efficiency aircraft structures utilize steel, aluminium or other metals, although reinforced plastics are sometimes adopted in special circumstances. In the building industry the choice of face material is much wider, including plywood, hardboard, plaster, plastics, asbestos cement and a host of composite materials.

If meaningful calculations are to be performed for the analysis and design of sandwich panels it is obviously necessary to have reliable information about the strength and stiffness of the materials used. It is not usually difficult to obtain the properties of common steels and aluminium alloys from appropriate works of reference, but the properties of other face materials are less easily discovered. For this reason it is often necessary to resort to tests to establish these properties. This is especially true in the case of materials which are not ordinarily used for sructural purposes, for which no special effort is made by the manufacturer to achieve uniform strength and stiffness.

In any efficient sandwich the faces act principally in direct tension and compression. It is therefore appropriate to determine the modulus of elasticity, ultimate strength and yield or proof stress of the face material in a simple tension test. If it is suspected that the properties in compression are different from those in tension it is desirable to repeat the tests in compression. Compression tests on thin sheet materials present considerable difficulty. If the specimen is small but fairly thick it may be possible to complete the test without premature failure due to elastic instability. Otherwise some form of lateral support is necessary. In one standard test[15.7] the specimen is clamped between two blocks of steel, the faces of which are grooved in the direction of the load in an attempt to permit the specimen to contract freely. In another,[15.8] intended originally for wood veneers, the specimen is supported laterally by the ends of long leaf-springs, set perpendicular to the specimen. Difficulty is sometimes caused by premature splitting of the ends of specimens if they are not clamped in some way.

When the material is thick and it is to be used with a weak core it may be desirable to determine its flexural rigidity. For homogeneous materials this is easily obtained by measurements of the thickness and of the modulus of elasticity in tension. For other materials such as reinforced plastics and plywood this may not provide very accurate results, in which case the flexural rigidity must be obtained by direct measurement. This is usually done by means of the four-point load test illustrated in Fig. 12.1a. A narrow strip of face material is supported at two points B, F and equal loads W are applied at A, G. The distances AB, FG are each equal to b and the central region is therefore subjected to a uniform bending moment Wb. As a result it bends in an arc of a circle of curvature $1/R = Wb/EI$. If the deflection of D with respect to the line CE is measured and denoted by Δ, it is easy to show that for small deflections,

$$EI = \frac{Wbc^2}{2\Delta}, \qquad (12.1)$$

where $CD = DE = c$. It is best not to rely on one measurement of Δ but to plot W against Δ for a range of loads and to insert in equation (12.1) the slope W/Δ.

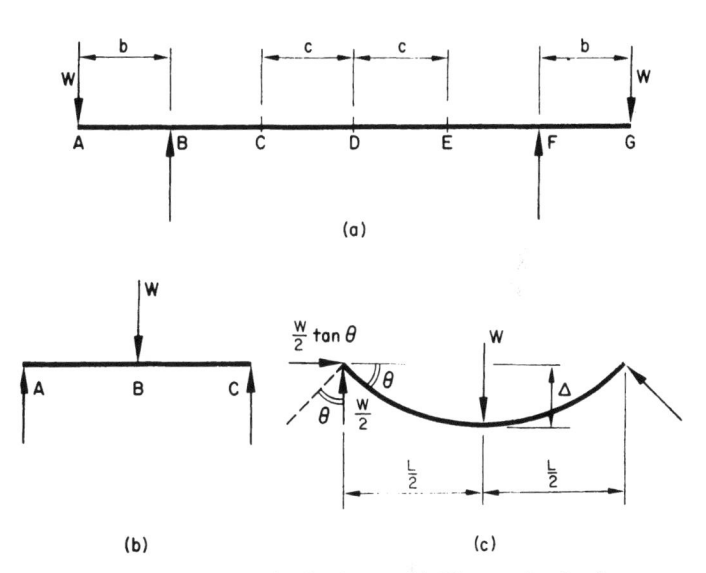

FIG. 12.1. (a) Four-point load test. (b) Three-point load test. (c) Effect of excessive deflection.

The three-point load test (Fig. 12.1b) may also be used in various ways to determine the flexural rigidity EI. However, this test is generally considered to be less accurate than the four-point load test. In both tests it is essential that deflections should not be so large that the horizontal component of the inclined reactions introduces a significant bending moment. For example, in Fig. 12.1c the central bending moment is not $WL/4$ but

$$\left(\frac{WL}{4} + \frac{W\Delta}{2} \tan \theta\right).$$

12.2. Core Materials and Their Properties

A core material is required to perform two essential tasks; it must keep the faces the correct distance apart and it must not allow one face to slide over the other. In other words it must be rigid and strong in direct tension and compression (perpendicular to the faces) and in shear (in the planes perpendicular to the faces). It must also be of low density. The properties of some typical sandwich materials are listed in Table 12.1 and the variation of the shear modulus G_{LT} with density is plotted in Fig. 12.2.

The suffixes L, W, T may be interpreted as "length", "width" and "transverse direction" respectively as shown in Fig. 12.3. This notation is used in refs. 11.9 and 12.4, but it should not be confused with an alternative notation T, R, L representing "tangential", "radial" and "longitudinal". These terms were used for balsa in refs. 13.1 and subsequently for honeycomb in refs. 11.4 and 11.5. The directions T, R, L correspond to L, W, T respectively.

Modern expanded plastics (group No. 1 in the table) are approximately isotropic and their strengths and stiffnesses are very roughly proportional to density. They are available in various densities and they may vary considerably in texture and characteristics from one manufacturer to another. All are rather weak in relation to their density (G_{LT} is too small to be plotted in Fig. 12.2) and all of them tend to soften at moderately elevated temperatures; phenolic foam is the most resistant to flame but it suffers from brittleness and friability. As a group, expanded plastics are unlikely to find applications in high-efficiency sandwich structures, but applications in semi-structural building panels with an insulating function are possible. Further information about the properties and manufacture of expanded plastics is given in refs. 15.15 and 15.16.

Some test results for aluminium alloy honeycomb are shown

in group 2 in Table 12.1. The $1/4$-in. cell size (Fig. 12.3b) is a fairly common one and it is evident that all the properties increase progressively with increases in thickness of the foil from which the honeycomb is made. The shear modulus G_{LT} is plotted in Fig. 12.2 which also shows values from other tests not listed in

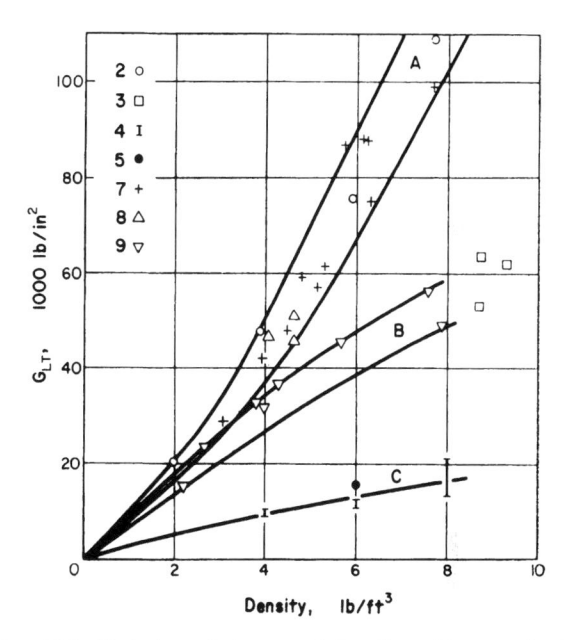

FIG. 12.2. Variation of shear modulus G_{LT} with core density for various cores. Nos. 2–5 correspond to groups 2–5 in Table 12.1. Nos. 7–9 are described in the text.

Table 12.1. For example, points marked 7 are for cores with cell sizes from $1/8$ in. to $3/8$ in. and various foils, manufactured by different processes.[11.5] Points marked 8 are for multiwave cores with $1/4$ in. cells.[11.6] It is remarkable that all of these points for aluminium honeycomb cores of different types fall within the narrow band marked A. All of these figures are the results of tests by the U.S. Forest Products Laboratory. It is no less interesting that the points marked 9, which fall into a different band B, represent values quoted by a manufacturer. Whether the difference between bands A and B is accounted for by differences in test methods or by real differences in the properties of the cores is a matter for conjecture. Nevertheless, it serves as a warning against the uncritical acceptance of figures stated in the literature.

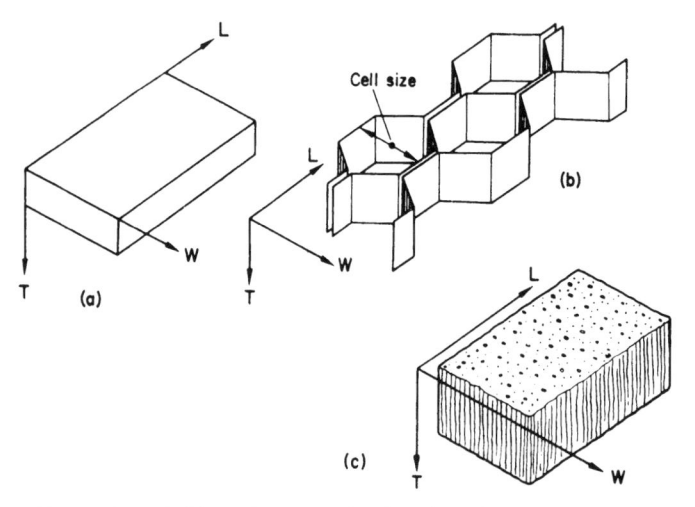

FIG. 12.3. Definition of L, W and T directions for core materials. (a) Slab of expanded plastics. (b) Honeycomb. (c) End-grain balsa.

Some selected test results for reinforced plastics are shown in Table 12.1 and Fig. 12.2. In the diagram, points marked 4 fall near the curve C but the points marked 3 are much higher. The reasons for the differences between groups 3 (ref. 11.9) and 4 (ref. 11.4) are not made clear in the original sources but it is obvious that great variations can be expected in the properties of plastics honeycomb which are apparently of similar construction.

MATERIALS USED IN SANDWICH CONSTRUCTION 251

TABLE 12.1. PROPERTIES OF CORE MATERIALS

E_T, σ_T are the modulus of elasticity and ultimate strength in compression in the T-direction (Fig. 12.3). G, τ are the shear modulus and ultimate shear strength in the LT- and TW-planes. Moduli, stresses and densities are in units of 1000 lb/in², lb/in² and lb/ft³ respectively. The figures quoted are generally mean values of results obtained by experiment.

Material	E_T	σ_T	G_{LT}	τ_{LT}	G_{TW}	τ_{TW}	Density
(1) Expanded plastics[15.18]							
PVC	8	150	1·35	165	(As for G_{LT}, τ_{LT})		4·1
Phenolic	1	25	0·5	20			3·5
Polyurethane	2·7	75	0·7	85			4·5
Polystyrene	1·6	40	0·65	45			2·6
(2) U.S. commercial aluminium honeycomb, $1/4$ in. cell size[11.9]							
0·0007-in. foil	–	145	20·4	136	10·2	79	1·99
0·002-in. foil	259	477	47·7	342	18·7	166	3·92
0·003-in. foil	437	899	75·5	565	25·7	276	5·88
0·004-in. foil	535	1386	109	771	32·3	383	7·70
(3) Reinforced plastics honeycomb, $3/16$-in. cell size[11.9]							
Phenolic/glass	179	1670	63·6	618	33·8	464	8·77
Phenolic/asbestos	198	1562	61·9	562	31·5	451	9·30
Silicone/asbestos	146	704	52·6	451	21·2	265	8·70
(4) Reinforced plastics honeycomb, $1/4$-in. cell size[11.4]							
max	54	460	9·8	180	4·4	100	} 4·0
min	36	350	8·8	175	2·8	63	
max	83	980	12·3	215	8·4	170	} 6·0
min	73	830	11·0	165	6·8	155	
max	104	1240	20·8	230	12·2	185	} 8·0
min	57	734	13·5	150	7·8	130	

TABLE 12.1 (cont.).

Material	E_T	σ_T	G_{LT}	τ_{LT}	G_{TW}	τ_{TW}	Density
(5) Balsa wood[13.1]							
	350	900	15	–	15	–	6
(6) Other materials[13.1, 13.4]							
Calcium alginate	12	140	4·9	–	–	–	6
Expanded formvar	6	250	3·5	–	–	–	6
Balsolite	4·5	192	2·1	–	–	–	4

More detailed information about reinforced plastics honeycombs is given in ref. 11.7, which describes tests on materials with different resins and glass reinforcements and with cell sizes of $\frac{3}{16}$ in. and $\frac{1}{4}$ in. The results for the tests under dry conditions are rather better than the values in group 4 in Table 12.1, but not so good as those in group 3. The report also gives some indication of the effect of moisture.

Details of steel, nickel–chromium and titanium honeycombs for heat-resisting structures are available in ref. 11.8. The variations in the results are considerable, depending on the nature of the metal and the geometry of the honeycomb and it is not possible to provide a concise numerical summary. Most of the cores have densities in the range 7·0 to 9·5 lb/ft^3. In comparison with aluminium alloy cores of similar density (e.g. Fig. 12.2, band A) they tend to have comparable stiffnesses but slightly lower strengths at normal temperatures.

Balsa wood is one of the original core materials. It is usually used with the grain perpendicular to the faces of the sandwich ("end-grain" balsa). The density is rather variable but the transverse strength and stiffness are good and the shear stiffnesses moderate (Table 12.1). Moduli in the L and W directions are less than 5% of the value quoted for the T-direction, but this is of little practical importance. The point for balsa (No. 5, Fig. 12.2) falls near the curve C.

Some materials of historical interest are listed in group 6, Table 12.1. Calcium alginate is rather brittle and expanded formvar (polyvinyl formal) softens at temperatures near 50°C. Balsolite is an early honeycomb made of closely-packed resin-impregnated paper tubes. Information about the manufacture of these and other materials is given by Topp.[13.2]

The properties of corrugated cores may be calculated from the formulae and graphs prepared by Libove and Hubka[12.4] Table 12.2 shows the results of calculations for some representative corrugated cores of aluminium alloy. The approximate value of G has been obtained by dividing the shear stiffness D_Q by d.

TABLE 12.2

$\theta = 75°; \quad p = 2d_c \quad$ (Fig. 7.3)

d_c/t_c	5	7.5	10	15	20	30	40	70
Density lb/ft³	50.9	35.9	27.8	19.1	14.5	9.9	7.5	4.3
G, 1000 lb/in² $t_c/t = 0.3$	264	77	32	8.8	3.6	1.0	0.41	0.074
$t_c/t = 1.25$	193	55	22	5.9	2.3	0.67	0.28	0.050

Comparison of the results in Table 12.2 with those in Fig. 12.2 shows that for any given density a honeycomb core is much stiffer than a corrugated core (in the plane perpendicular to the corrugations). This must be balanced against the fact that the corrugated core has a very high shear stiffness in the direction of the corrugations; for practical calculations it is usually assumed to be infinite.

Theories are available for the prediction of the properties of honeycomb cores. Kelsey, Gellatly and Clark[15.10] give a thorough analysis which provides upper and lower bounds for the shear modulus. A less comprehensive analysis is given by Chang and Ebcioglu[15.14] and yet another by Penzien and Didriksson.[15.17] The value of Poisson's ratio for honeycomb cores is rather uncertain and a note on the subject is offered by Hoffman.[14.1]

Miscellaneous information about the properties of core materials will be found in refs. 5.12, 11.3, 11.10, 12.1, 12.2, 12.3, 12.5, 15.9, 15.11, 30.7, 35.5 and 35.10.

12.3. Methods of Testing Core Materials

Although Table 12.1 gives a useful indication of the relative properties of different kinds of core material, no such general table can be sufficiently precise for design purposes. It is therefore nearly always necessary to determine the strength and stiffness of any particular form of core material by experiment. A number of tests have been devised over the years and some of them have been enshrined in standard specifications.

Ref. 15.6 gives details of a test for the determination of the tensile strength of the core in the T-direction; alternatively, the test provides the tensile strength of the bond between the faces and the core. A sample of the sandwich at least 1×1 in. in the LW-plane is glued between two metal blocks which are pulled apart by a self-aligning loading rig. Larger samples may be necessary for cores with large cell sizes.

Recommendations for the measurement of compressive strength in the T-direction are given in ref. 15.2, which also gives instructions for the measurement of strains and hence the modulus E_T. It is considered essential to use an extensometer to measure the strain in the core directly, rather than to compute the strain from measurement of the relative movements of the crossheads of the testing machine.

The strength of the adhesive between the faces and the core may be measured by the climbing-drum peel test.^(15.4)

Much attention has been given in the literature to the selection of the best method of measuring the shear strength and stiffness in the *LT*-and *TW*-planes. The most obvious method is the use of the single- or double-block shear test (Fig. 12.4a, b). Samples of the core *C* (or of the complete sandwich) are glued to thick

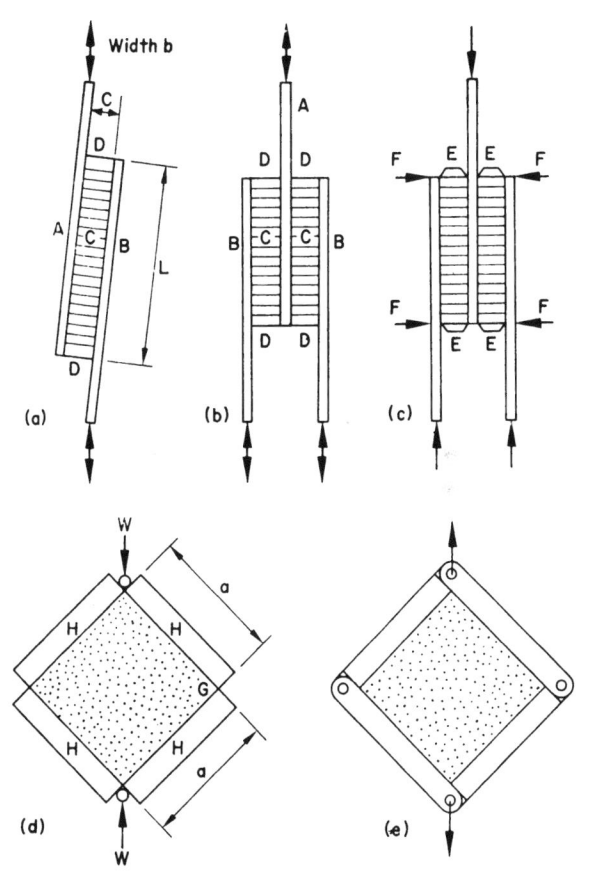

FIG. 12.4. Types of shear test.

metal faces A, B and loaded in the directions shown. The shear stress is merely the load divided by the surface area of the specimen and the shear strain is the relative movement of the faces A and B divided by the thickness of the core specimen. For the dimensions shown in Fig. 12.4a the shear modulus is $cW \cos \theta / bL\Delta$, where Δ is the relative displacement of the faces along their lengths. Details of a single-block shear test rig are given in ref. 15.5.

Some minor criticisms of these two tests are made by Kelsey et al.[15.10] who consider the effects of non-uniform stress in the face plates (both types), bending of the face plates and a tensile component of the load across the specimen (single-block) or the action of the rig as a deep beam of short span (double-block). Possibly a more important criticism is the omission of any provision for carrying the complementary shear stress at the ends of the specimen (D in Fig. 12.4a, b). Because this edge is completely free, the shear stress there must be zero and the specimen cannot be in the assumed state of uniform shear. O'Sullivan[15.13] blames this feature for the tendency of the double-block shear test to give unduly low values for G. He describes a modified shear test (Fig. 12.4c) in which "bridge pieces" E are attached to the ends of the core. These bridge pieces are intended to transmit the complementary shear stresses to the face plates and, in order to ensure that proper contact is made, clamping forces F are introduced. It is claimed that this procedure produces results which are in good agreement with the theory of Kelsey.[15.10]

A simple shear test which presents few problems for a material which is available in bulk is shown in Fig. 12.4d. A square specimen is glued to four rigid blocks H and loaded in compression through two small rollers which bear against the ends of the blocks. If the dimensions of the block are $a \times a \times t$, the shear stress is $W/at\sqrt{2}$. The shear strain γ is equal to $2 e_c$ or $2 e_t$ where e_c is the direct compressive strain in the vertical direction, and e_t is the direct tensile strain in the horizontal direction. The shear modulus is therefore $W/at_e 2\sqrt{2}$. A rather similar test is described by Penzien and Didriksson[15.17] but in their version the blocks are hinged together to form a racking frame which is loaded in tension (Fig. 12.4e). The method is not well adapted for the determination of properties in the LT- and TW-planes when the core material is available only in sheet form rather than in bulk.

Further information about these methods of testing core materials can be found in refs. 11.2 and 13.3. Torsion tests for the determination of the core shear modulus are described briefly in refs. 11.1 and 13.3.

12.4. Tests on Sandwich Constructions

In the chapter on wrinkling it was emphasized that any new form of sandwich construction should be subjected to tests to verify the applicability of the formulae for wrinkling, tensile bond failure or intracellular buckling of the faces. Such tests are easily carried out by subjecting a short sample of the complete sandwich to a compressive load in the L- or W-direction. Ref. 15.1 gives details of an appropriate test rig in which the sandwich is loaded as a short pin-ended strut.

It is often desired to determine the flexural and shear rigidities of a particular sample of sandwich construction and the three-point load test (Fig. 12.1b) is commonly used for this purpose. It was shown in Section 2.3 that the deflection Δ under a central point load W is

$$\Delta = \Delta_1 + \Delta_2 = \frac{WL^3}{48D} + \frac{WL}{4AG}, \tag{12.2}$$

where D is the flexural rigidity, AG is the shear stiffness and L is the span. For most purposes G is the core shear modulus and A is equal to bd^2/c (Fig. 2.1). It appears that if the deflection is measured at two different spans, the resulting pair of equations can be solved for D and AG. This is unlikely to lead to good results unless it is known that one span is large enough to ensure that the deflection is mostly due to bending and the other is small enough to ensure that the deflection is mostly due to shear. This is not always easy and a more reliable method is as follows.

Equation (12.2) can be recast in two other forms:

$$\frac{\Delta}{WL} = \frac{L^2}{48D} + \frac{1}{4AG}, \qquad (12.3a)$$

$$\frac{\Delta}{WL^3} = \frac{1}{48D} + \frac{1}{4AG} \cdot \frac{1}{L^2}. \qquad (12.3b)$$

The first of these equations can be represented as a straight line in a plot of Δ/WL against L^2, as shown by the full line in Fig. 12.5a. The second can be represented as a straight line in a plot

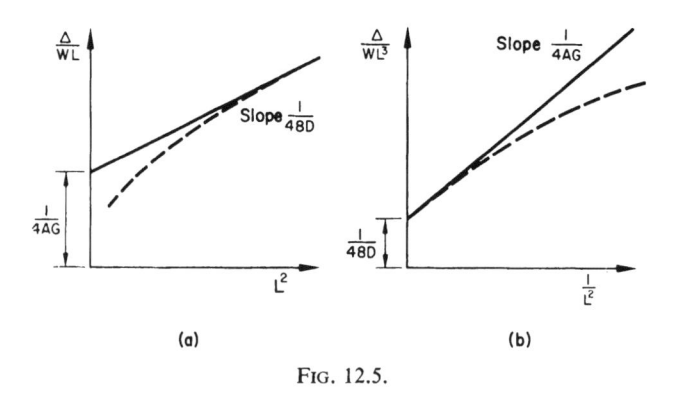

FIG. 12.5.

of Δ/WL^3 against $1/L^2$, as shown by the full line in Fig. 12.5b. If the flexibility Δ/W of the beam is measured at a number of different spans, the straight lines in Fig. 12.5 may be plotted and the required stiffnesses D and AG may be obtained from the slopes and from the intercepts on the vertical axes. If the flexural rigidity D is already known as a result of other tests, it is convenient to use it to mark the intercept in Fig. 12.5b and then to determine AG from the slope of the line in that diagram.

Precautions must be observed in the evaluation of the flexibility Δ/W at a given span. It is desirable to measure the deflections at a number of different loads, to plot Δ against W and to obtain Δ/W from the slope of the resulting straight line. The central deflection should be measured on the *underside* of the beam at B (Fig. 12.1b). The settlement over the support should be measured at the top face of the beam at points A and C, and the mean value should be subtracted from the central deflection. In this way the effect of the local crushing of the core at A, B and C is minimized.

The method gives good results except for sandwiches with thick faces and very weak cores; a sandwich with plywood faces and an expanded polystyrene core might fall into this category. In these extreme cases the plotted results follow the curves indicated by the broken lines in Figs. 12.5. Allen[20.3] shows that this departure from the simple theory is due to the modification of the shear deflection by the local bending stiffness of the faces. The effect was discussed in detail in Section 2.5 et seq. and the deflection under a central point load is given in general by equation (2.37a), which may be written in a form comparable with equation (12.3b) as follows:

$$\frac{\Delta}{WL^3} = \frac{1}{48D} + \frac{1}{4AG}\left\{1 - \frac{I_f}{I}\right\}^2 \psi_1 \left\{\frac{1}{L^2}\right\}, \quad (12.4)$$

where $\qquad I = \dfrac{bt^3}{6} + \dfrac{btd^2}{2} \quad$ and $\quad I = \dfrac{bt^3}{6}.$

The coefficient ψ_1 depends on the overhang at the ends of the beam, the proportions of the sandwich and the ratio of the moduli of the face and core materials, in a manner described in Sections 2.5 and 2.6. It is sufficient to note that ψ_1 tends to unity as $1/L^2$ tends to zero. Equation (12.4) therefore indicates that the slope of the left-hand end of the curve in Fig. 12.5b is equal to

$$\frac{1}{4AG}\left(1-\frac{I_f}{I}\right)^2$$

and this fact may be used to determine AG. The result is approximate, however, and better results can only be obtained by fitting equation (12.4) to the plotted points by trial and error.

In any particular test a straight-line plot of the results can be expected if ψ_1 is close to unity; Fig. 2.10 shows that this is true for values of θ greater than about 15. The value of θ may be obtained from Fig. 2.11 or equation (2.39). In general there is a minimum span for which the simple equations (12.3) can be used in place of the elaborate equation (12.4); further details are given in ref. 20.3.

Another analysis of the three-point load test as applied to beams with honeycomb cores is presented by Cox and Martin.[20.2]

The four-point load test illustrated in Fig. 12.1a may be used to determine the flexural rigidity of a narrow sandwich beam. Equation (12.1) can be used and the result is the total flexural rigidity as defined by equation (2.2). Although the central region BF of the beam is in a state of pure bending, it is still possible for the results to be misleading when the core is very weak in shear and the faces are thick. For example, if the shear stiffness of the core is so small as to be negligible, the two faces bend as independent beams (Fig. 2.16b) and the application of equation (12.1) results in the evaluation of the term $E_f bt^3/6$ rather than $(E_f bt^3/6 + E_f btd^2/2)$. The magnitude of the error involved is discussed in Section 2.8 but for simplicity it is probably safe to assume that it is small provided the value of ψ_1 in equation (12.4) is close to unity at the span used (ψ_1 can be found fairly quickly from Figs. 2.10 and 2.11).

Howard[20.1] describes various methods of finding the flexural and shear rigidities from the five-point load test (Fig. 12.6). The object is to use the loads W to produce a bending deflection which counteracts the bending deflection due to $2F$, leaving only the shear deflection due to $2F$ to be measured.

Kuenzi[16.4] gives a straightforward description of the use of the four-point load test and a detailed specification is given in

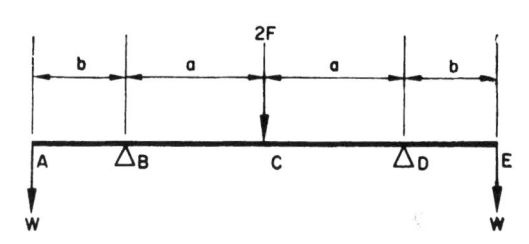

FIG. 12.6. Five-point load test.

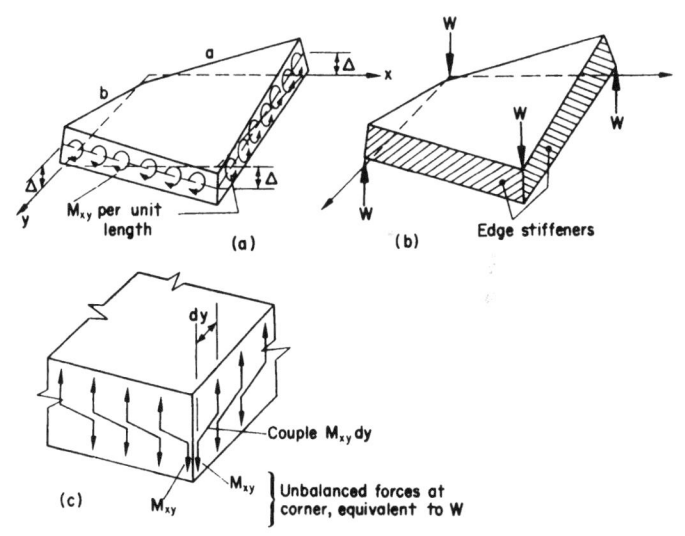

FIG. 12.7. Torsion test on sandwich panel. (a) Ideal loading by edge twisting couples. (b) Practical loading by point loads and edge stiffeners. (c) Equivalence of ideal and practical loadings.

ref. 15.3, which describes the evaluation of core shear strength, strength of the faces, flexural rigidity and shear stiffness. The last two are obtained by the solution of two simultaneous equations formed from deflection measurements at two different spans. These equations, and those of Howard, are not valid when the faces are exceptionally thick and the core is weak. Useful descriptions of tests on sandwich beams are given in refs. 16.3, 19.1.

The twisting stiffness D_{xy} of a plate is defined by equation (7.3) and its value can be found by applying twisting couples M_{xy} per unit length to the edges of the plate as in Fig. 12.7a. The slopes $\partial w/\partial y$ in the y-direction are $-2\varDelta/b$ at $x = 0$ and $+2\varDelta/b$ at $x = a$, where $\varDelta$ is the displacement at each corner. The rate of change of $\partial w/\partial y$ in the x-direction is therefore

$$\frac{\partial^2 w}{\partial x\, \partial y} = \left\{\left(\frac{2\varDelta}{b}\right) - \left(-\frac{2\varDelta}{b}\right)\right\}\bigg/a = \frac{4\varDelta}{ab}. \qquad (12.5)$$

The most convenient way of applying the twisting couples M_{xy} is to fit thin stiffeners to each of the four edges of the panel and to load the panel at the corners as in Fig. 12.7b. The twisting couples near the corner ($x = a$, $y = b$) can be represented as in Fig. 12.7c, from which it is clear that

$$W = 2M_{xy}. \qquad (12.6)$$

From equations (7.3), (12.5) and (12.6), therefore, the twisting stiffness is given by

$$D_{xy} = \frac{Wab}{8\varDelta}. \qquad (12.7)$$

Although this test has been proposed in several papers there are few detailed descriptions. Some practical information is given in refs. 11.2, 12.4.

Measurements of the effective values of Poisson's ratios for sandwich panels have not often been attempted. Theoretically it should be possible to subject a panel to pure anticlastic bending and to determine the Poisson's ratios from the ratios of the curvatures in different directions, in accordance with equations (7.1b) and (7.2b). A less direct method which is possibly less reliable and which is only appropriate for isotropic sandwiches is to determine $D_x (= D_y)$ and D_{xy} from bending and twisting experiments and to calculate Poisson's ratio from Section 7.6 (i), viz.

$$v = \frac{D_x}{D_{xy}} - 1 = \frac{D_y}{D_{xy}} - 1. \tag{12.8}$$

In determining D_x it is essential that the panel is bent anticlastically, not cylindrically; this can be done by testing a narrow strip of the sandwich as a beam.

Equations (12.7) and (12.8) should not be used when the faces are very thick and the core is very weak.

APPENDIX I

PROPERTIES OF ISOTROPIC AND ORTHOTROPIC ELASTIC SOLIDS

IN AN orthotropic elastic solid it is always possible to find three mutually perpendicular axes (x, y, z) such that when a direct stress σ_x is applied, pure extensions occur in the x, y and z-directions, while the shear strains in the zx-, xy-, yz-planes are zero. Throughout this book it has been assumed that orthotropic face and core materials are orientated in this way with respect to the main axes of the beam or panel under consideration.

The direct strains e_x, e_y and e_z are related to the stress σ_x in the following manner:

$$e_x = \frac{\sigma_x}{E_x}; \quad e_y = -\nu_{xy}e_x = -\frac{\nu_{xy}\sigma_x}{E_x};$$

$$e_z = -\nu_{xz}e_x = -\frac{\nu_{xz}\sigma_x}{E_x}. \quad (\text{I}.1)$$

E_x is the modulus of elasticity in the x-direction; ν_{xy} and ν_{xz} are the Poisson's ratios associated with stress in the x-direction and strains in the y- and z-directions. The negative signs arise because tensile stresses and extensional strains are both positive. Equation (I.1) constitutes a definition of the Poisson's ratios ν_{xy}, ν_{xz}.

Similar equations may be written down for the strains associated with stresses in the y- and z-directions in turn. All three equations may be superimposed to express the direct strains associated with stresses σ_x, σ_y and σ_z applied simultaneously:

$$e_x = \frac{\sigma_x}{E_x} - \nu_{yx}\frac{\sigma_y}{E_y} - \nu_{zx}\frac{\sigma_z}{E_z}, \tag{I.2}$$

$$e_y = -\nu_{xy}\frac{\sigma_x}{E_x} + \frac{\sigma_y}{E_y} - \nu_{zy}\frac{\sigma_z}{E_z}, \tag{I.3}$$

$$e_z = -\nu_{xz}\frac{\sigma_x}{E_x} - \nu_{yz}\frac{\sigma_y}{E_y} + \frac{\sigma_z}{E_z}. \tag{I.4}$$

The shear strains and stresses are related as follows:

$$\tau_{zx} = G_{zx}\gamma_{zx}; \quad \tau_{xy} = G_{xy}\gamma_{xy}; \quad \tau_{yz} = G_{yz}\gamma_{yz}. \quad (I.5, I.6, I.7)$$

Equations (I.2)–(I.7) describe the elastic behaviour of a general orthotropic elastic solid. They would be employed, for example, in any study of the properties of an orthotropic core material.[16.3]

In orthotropic face materials some simplification is possible because the stress σ_z is generally neglected, as are the shear strains γ_{zx} and γ_{yz}. If these items are equated to zero, equations (I.5) and (I.7) disappear, equation (I.4) is not usually needed, leaving the following:

$$e_x = \frac{\sigma_x}{E_x} - \nu_{yx}\frac{\sigma_y}{E_y}, \tag{I.8}$$

$$e_y = -\nu_{xy}\frac{\sigma_x}{E_x} + \frac{\sigma_y}{E_y}, \tag{I.9}$$

$$\tau_{xy} = G_{xy}\gamma_{xy}. \tag{I.10}$$

Equations (I.8) and (I.9) may be rearranged as follows:

$$\sigma_x = \frac{E_x}{g}(e_x + \nu_{yx}e_y), \tag{I.11}$$

$$\sigma_y = \frac{E_y}{g}(e_y + \nu_{xy}e_x), \quad \text{where} \quad g = 1 - \nu_{xy}\nu_{yx}. \tag{I.12}$$

Equations (I.10)–(I.12) now describe the properties of the orthotropic face of a sandwich plate.

The strain energy of an orthotropic elastic solid is equal to the following integral:

$$U = \frac{1}{2}\int_V (\sigma_x e_x + \sigma_y e_y + \sigma_z e_z + \tau_{zx}\gamma_{zx} + \tau_{xy}\gamma_{xy} + \tau_{yz}\gamma_{yz})dV, \quad (I.13)$$

where the integration is carried out over the volume of the solid. For the orthotropic face of a sandwich panel it has already been noted that σ_z, γ_{zx} and γ_{yz} may be equated to zero. If, in addition, the stresses σ_x, σ_y and τ_{xy} are expressed in terms of the strains by the use of equations (I.10)–(I.12), the strain energy appears as follows:

$$U = \frac{1}{2}\int_V \{E_x e_x^2/g + E_y e_y^2/g + 2E_x \nu_{yx} e_x e_y/g + G_{xy}\gamma_{xy}^2\}\, dV. \quad (I.14)$$

In the derivation of this equation use has been made of the fact that $E_x\nu_{yx}$ and $E_y\nu_{xy}$ are equal, which follows from the Reciprocal Theorem. It should be noted that in general G_{xy} cannot be expressed in terms of the elastic moduli and the Poisson's ratios; that is possible only when the material is isotropic.

A beam which is loaded in the zx-plane can deform in one of two ways: anticlastic bending, or cylindrical bending. In anticlastic bending the beam is free to expand or contract laterally; thus $\sigma_y = 0$ in equations (I.8) and (I.9), which become, simply:

$$\sigma_x = E_x e_x, \quad (I.15)$$

$$e_y = -\nu_{xy} e_x. \quad (I.16)$$

Equation (I.15) has been used in Chapters 2 and 3 to describe the behaviour of the faces of sandwich beams and struts.

In cylindrical bending the beam is constrained so that there is no strain in the y-direction; thus $e_y = 0$ in equations (I.11) and (I.12), which become:

$$\sigma_x = \frac{E_x}{g} e_x, \tag{I.17}$$

$$\sigma_y = \nu_{xy} \frac{E_y}{E_x} \sigma_x. \tag{I.18}$$

Thus, for cylindrical bending equation (I.17) should be used instead of (I.15). To convert from anticlastic bending to cylindrical bending in Chapters 2 and 3 it is therefore necessary to replace E_x everywhere by E_x/g.

Anticlastic bending is generally considered to occur when a beam is narrow in relation to its depth. Cylindrical bending occurs when the beam is wide and especially if the deflections are large and if constraints are present which inhibit curvature in the yz-plane.

All of the formulae given here are appropriate for isotropic solids, for which $E_x = E_y = E_z$, $\nu_{xy} = \nu_{yx} = \nu_{xz} = \nu_{zx} = \nu_{yz} = \nu_{zy}$. In addition, $G_{zx} = G_{xy} = G_{yz} = E/2(1+\nu)$.

APPENDIX II

DIFFERENTIAL EQUATION FOR A SANDWICH BEAM-COLUMN

IN CHAPTER 2 three equations were constructed to describe the behaviour of a sandwich beam with thick faces. These equations are listed below, with certain modifications to convert them from anticlastic to cylindrical bending and to bring the notation into line with that of Chapter 7, on plates.

$$Q_1 = -(D_2 + D_f) w_1''', \qquad (2.18)$$

$$Q_1 = a^2 D_f w_2', \qquad (2.28)$$

$$Q_1'' - a^2 Q_1 = -a^2 Q, \qquad (2.27a)$$

where

$$a^2 = \frac{D_Q}{D_f \{1 - D_f/(D_2 + D_f)\}} \qquad (2.27b)$$

and $D_Q = Gd^2/c$; $D_f = Ebt^3/6g$; $D_2 = Ebd^2t/2g$; $g = (1-v^2)$.

It will be recollected that w_1 is the displacement in the absence of shear strain in the core, Q_1 is the total shear force carried by the beam in association with the displacement w_1, Q is the shear force actually applied to the beam and w_2 is the additional displacement due to shear strain in the core (corresponding to Q_1). The total displacement is $w = w_1 + w_2$.

If the member supports an axial thrust P, the shear force Q in equation (2.27a) must be replaced by the sum of $P(w_1' + w_2')$ (the shear force due to P acting at an angle to the deflected centre line) and Q (the shear force due to transverse loads q acting alone).

Equation (2.27a) therefore appears as follows:

$$Q_1'' - a^2 Q_1 = -a^2 P(w_1' + w_2') - a^2 Q. \tag{II.1}$$

The value of Q_1 may be substituted from (2.18) into (II.1). After differentiation with respect to x there remains:

$$(D_2 + D_f)(w_1^{\text{vi}} - a^2 w_1^{\text{iv}}) = a^2 P(w_1'' + w_2'') - a^2 q. \tag{II.2}$$

The value of Q_1 may also be substituted from equation (2.28) into (II.1). After differentiation with respect to x there remains:

$$(D_2 + D_f)(w_2^{\text{vi}} - a^2 w_2^{\text{iv}}) = -P(w_1^{\text{iv}} + w_2^{\text{iv}})\left(1 + \frac{D_2}{D_f}\right) + q''\left(1 + \frac{D_2}{D_f}\right). \tag{II.3}$$

The addition of (II.2) and (II.3) provides a differential equation for w in terms of the axial and transverse loads P and q:

$$(D_2 + D_f)(w^{\text{vi}} - a^2 w^{\text{iv}}) = a^2 P w'' - P w^{\text{iv}}\left(1 + \frac{D_2}{D_f}\right) - a^2 q + q''\left(1 + \frac{D_2}{D_f}\right). \tag{II.4}$$

Or,

$$-\frac{D_f D_2}{D_Q} w^{\text{vi}} + (D_2 + D_f) w^{\text{iv}} = \left(1 - \frac{D_2}{D_Q}\frac{d^2}{dx^2}\right)(q - P w''). \tag{II.5}$$

This is the two-dimensional analogue of equation (9.1) for a sandwich plate. It has a certain interest as such, but for practical purposes the methods of Sections 2.6 and 3.1 may be found more convenient.

REFERENCES

REFERENCES are grouped according to subject. Within each group there are separate subdivisions for items from the U.S. Forest Products Laboratory (F.P.L.), the N.A.C.A. (or N.A.S.A.), the U.K. Aeronautical Research Committee (A.R.C), the *Journal of Aeronautical (or Aerospace) Sciences* and for items from other sources. Within each subdivision references are listed in order of original reference number or publication date, with the exception of refs.15.1–15.8.

Bending and Buckling of Sandwich Panels

1.1. H. W. MARCH, *Buckling Loads of Panels having Light Cores and Dense Faces*, FPL Report 1504, Feb. 1944.
1.2. H. W. MARCH and C. B. SMITH, *Buckling Loads of Flat Sandwich Panels in Compression. Various Types of Edge Conditions*, FPL Report 1525, Mar. 1945.
1.3. K. H. BOLLER, *Buckling Loads of Flat Sandwich Panels in Compression*, FPL Reports 1525-A, Feb. 1947; 1525-B, C, D, Sept. 1947; and 1525-E, Mar. 1948.
1.4. E. W. KUENZI, W. S. ERICKSEN and J. J. ZAHN, *Shear Stability of Flat Panels of Sandwich Construction*, FPL Report 1560, May 1947, last revised May 1962.
1.5. H. W. MARCH, *Effects of Shear Deformation in the Core of a Flat Rectangular Sandwich Panel.* (1) *Buckling under compressive end load.* (2) *Deflection under transverse load*, FPL Report 1583, May 1948.
1.6. W. J. KOMMERS and C. B. NORRIS, *Effects of Shear Deformation in the Core of a Flat Rectangular Sandwich Panel. Stiffness of flat panels with uniformly distributed loads normal to their surfaces—simply-supported edges*, FPL Report 1583-A, Oct. 1948.
1.7. W. S. ERICKSEN and H. W. MARCH, *Compressive Buckling of Sandwich Panels having Dissimilar Faces of Unequal Thickness*, FPL Report 1583-B, Nov. 1950, revised Nov. 1958.
1.8. W. S. ERICKSEN, *Effects of Shear Deformation in the core of a Flat Rectangular Sandwich Panel. Deflection under uniform load of sandwich panels having faces of unequal thickness*, FPL Report 1583-C, Dec. 1950.
1.9. W. S. ERICKSEN, *Effects of Shear Deformation in the Core of a Flat*

Rectangular Sandwich Panel. Deflection under uniform load of sandwich panels having facings of moderate thickness, FPL Report 1583-D, Dec. 1951.
1.10. E. W. KUENZI, *Edgewise Compression Strength of Panels and Flatwise Flexural Strength of Strips of Sandwich Construction*, FPL Report 1827, Nov. 1951.
1.11. H. W. MARCH, *Behaviour of a Rectangular Sandwich Panel under a Uniform Lateral Load and Compressive Edge Load*, FPL Report 1834, 1952.
1.12. C. B. NORRIS and W. J. KOMMERS, *Stresses within a Rectangular, Flat Sandwich Panel subjected to a Uniformly Distributed Normal Load and Edgewise Direct and Shear Load*, FPL Report 1838, Jan. 1953.
1.13. M. E. RAVILLE, *Deflection and Stresses in a Uniformly Loaded, Simply-supported, Rectangular Sandwich Plate*, FPL Report 1847, Dec. 1955.
1.14. W. C. LEWIS, *Supplement to "Deflections and Stresses in a uniformly Loaded, Simply-supported, Rectangular Sandwich Plate"*, FPL Report 1847-A, Dec. 1956.
1.15. C. B. NORRIS, *Compressive Buckling Curves for Flat Sandwich Panels with Isotropic Facings and Isotropic or Orthotropic Cores*, FPL Report 1854, revised Jan. 1958.
1.16. C. B. NORRIS, *Compressive Buckling Curves for Simply-supported Sandwich Panels with Glass-fabric-laminate Facings and Honeycomb Cores*, FPL Report 1867, Dec. 1958.
1.17. E. W. KUENZI, C. B. NORRIS and P. M. JENKINSON, *Buckling Coefficients for Simply-supported and Clamped Flat Rectangular Sandwich Panels under Edgewise Compression*, U.S. Forest Service Research Note FPL-070, Dec. 1964.
1.18. P. M. JENKINSON, and E. W. KUENZI, *Buckling Coefficients for Flat Rectangular Sandwich Panels with Corrugated Cores under Edgewise Compression*. U.S. Forest Service Research Paper FPL-25, May 1965.
2.1. C. LIBOVE and S. B. BATDORF, *A General Small-deflection Theory for Flat Sandwich Plates*, NACA TN 1526, 1948. Also NACA Report 899.
2.2. P. SEIDE and E. Z. STOWELL, *Elastic and Plastic Buckling of Simply-supported Metalite Type Sandwich Plates in Compression*, NACA TN 1822, 1949. Also NACA Report 967.
2.3. P. SEIDE, *Compressive Buckling of Flat Rectangular Metalite Type Sandwich Plates with Simply-supported Loaded Edges and Clamped Unloaded Edges*, NACA TN 1886, 1949. Revised as NACA TN 2637, 1952.
2.4. P. SEIDE, *Shear Buckling of Infinitely Long Simply-supported Metalite Type Sandwich Plates*, NACA TN 1910, 1949.
2.5. N. J. HOFF, *Bending and Buckling of Rectangular Sandwich Plates*, NACA TN 2225, 1950.

REFERENCES

2.6. K. T. YEN, V. L. SALERNO and N. J. HOFF, *Buckling of Rectangular Sandwich Plates Subject to Edgewise Compression with Loaded Edges Simply-supported and unloaded Edges Clamped.* NACA TN 2556, 1952.

2.7. K. T. YEN, S. GUNTURKIN and F. V. POHLE, *Deflections of a Simply-supported Rectangular Sandwich Plate Subjected to Transverse Loads,* NACA TN 2581, 1951.

2.8. C. T. WANG, *Principle and Application of Complementary Energy Method for Thin Homogeneous and Sandwich Plates and Shells with Finite Deflections,* NACA TN 2620, 1952.

2.9. P. SEIDE, *The Stability under Longitudinal Compression of Flat symmetric Corrugated-core Sandwich Plates with Simply-supported Loaded Edges and Simply-supported or Clamped Unloaded Edges,* NACA TN 2679, 1952.

2.10. P. P. BIJLAARD, *Method of Split Rigidities and its Application to Various Buckling Problems,* NACA TN 4085, 1958.

3.1. D. M. A. LEGGETT and H. G. HOPKINS, *Sandwich Panels and Cylinders under Compressive End Loads,* A.R.C., R & M 2262 Aug. 1942.

3.2. H. G. HOPKINS and S. PEARSON, *The Behaviour of Flat Sandwich Panels under Uniform Transverse Load,* RAE Report SME 3277, 1944.

3.3. R. C. CHAPMAN, *Compression Tests on Dural-balsa Sandwich Panels,* A.R.C., R & M 2153, 1945.

3.4. W. S. HEMP, *On a Theory of Sandwich Construction,* A.R.C., R & M 2672, Mar. 1948.

3.5. J. H. HUNTER-TOD, *The Elastic Stability of Sandwich Plates,* A.R.C., R & M 2778, Mar. 1949.

4.1. E. REISSNER, Finite deflections of sandwich plates, *J. Aero. Sci.* **15,** 7, July 1948, pp. 435–40. Also ibid. **17,** 2, Feb. 1950, p. 125.

4.2. P. P. BIJLAARD, Analysis of elastic and plastic stability of sandwich plates by the method of split rigidities *J. Aero. Sci.* **18,** 5, and 12, May and Dec. 1951, pp. 339–49, 790–6 and 829. Also *ibid.* **19,** 7, July 1952, pp. 502–3.

4.3. G. GERARD, Note on bending of thick sandwich plates, *J. Aero. Sci.* **18,** 6, June 1951, pp. 424–7.

4.4. S. V. NARDO, An exact solution for the buckling load of flat sandwich panels with loaded edges clamped, *J. Aero. Sci.* **20,** 9, Sept. 1953, pp. 605–12.

4.5. G. A. THURSTON, Bending and buckling of clamped sandwich plates, *J. Aero. Sci.* **24,** 6, June 1957, pp. 407–12.

4.6. L. A. HARRIS and R. A. AUELMANN, Stability of flat simply-supported corrugated-core sandwich plates under combined loads, *J. Aero. Sci.* **27,** 7, July 1960, pp. 525–34. See also P. SEIDE, *ibid.,* **28,** 3, Mar. 1961, p. 248.

4.7. C. C. CHANG and B. T. FANG, Initially warped sandwich panels under combined loadings, *J. Aero. Sci.* **27,** 10, Oct. 1960, pp. 779–87.

REFERENCES

5.1. E. REISSNER, Effect of transverse shear deformation on the bending of elastic plates, *J. App. Mech.* **12**, 2. June 1945, pp. A69–77. See also *ibid.* **13**, 3, Sept. 1946, pp. A249–52.

5.2. E. REISSNER, On bending of elastic plates, *Quart. App. Math.* **5**, 1, Apr. 1947, pp. 55–68.

5.3. J. LOCKWOOD TAYLOR, Strength of sandwich panels, *Proc. VII Int. Congr. App. Mech.* **I**, 1948, pp. 187–99.

5.4. F. J. PLANTEMA, Some investigations on the Euler-instability of flat sandwich panels with simply-supported edges, *Proc. VII Int. Congr. App. Mech.* **I**, 1948, pp. 200–13.

5.5. A. C. ERINGEN, Bending and buckling of rectangular sandwich plates, *Proc. 1st U.S. Nat. Congr. App. Mech.* 1951, pp. 381–90.

5.6. G. GERARD, Linear bending theory of isotropic sandwich plates by an order-of-magnitude analysis, *J. App. Mech.* **19**, 1, Mar. 1952, pp. 13–15.

5.7. J. R. ROBINSON, The buckling and bending of orthotropic sandwich panels with all edges simply supported, *Aero. Quart.* **6**, 2, May 1955, pp. 125–48.

5.8. ANON., Information on use of data sheets on sandwich panels in compression, Roy. Aero. Soc. Data Sheet 07.01.00, *Structures* IV, 1956.

5.9. ANON., Buckling loads in compression of flat sandwich panels, Roy. Aero. Soc. Data Sheets 07.01.01, 02, 03, 04, *Structures* IV, 1956.

5.10. ANON., Transverse shear stiffness of corrugated-core sandwich panels with equal face plates, Roy. Aero. Soc. Data Sheets 07.03.01, *Structures* IV, 1956.

5.11. C. C. CHANG, B. T. FANG and I. K. EBCIOGLU, Elastic theory of a weak-core sandwich panel initially warped, simply supported and subjected to combined loadings *Proc. 3rd U.S. Nat. Congr. App. Mech.* 1958, pp. 273–80.

5.12. ANON., *Composite Construction for Flight Vechicles*, U.S. Military Handbook MIL-HDBK-23 Parts I, II, III, Oct. 1959 and later. Especially "Sandwich construction for aircraft" II and "Design procedures" III.

5.13. W. G. HEATH, Sandwich construction. Correlation and extension of existing theory of flat panels subjected to lengthwise compression, *Aircraft Engg.* **32**, 377 and 378, July and Aug. 1960, pp. 186–91 and 230–5.

5.14. C. C. CHANG and I. K. EBCIOGLU, Elastic instability of rectangular sandwich panels of orthotropic core with different face thicknesses and materials, *J. App. Mech.* **27**, 3, Sept. 1960, pp. 474–80.

5.15. SHUN CHENG, On the theory of bending of sandwich plates, *Proc. 4th U.S. Nat. Congr. App. Mech.* **I**, 1962, pp. 511–18.

5.16. W. R. KIMEL and M. E. RAVILLE, Buckling of sandwich panels in

edgewise bending and compression, *Proc. 4th U.S. Nat. Congr. App. Mech.* **I**, 1962, pp. 657–66.
5.17. A. M. ALWAN, Large deflection and small deflection analyses applied to rectangular sandwich plates with isotropic cores, Ph.D. Thesis, Dept. Engg. Mech., Univ. of Wisconsin, Madison, Wis., June 1963.
5.18. A. M. ALWAN, Large deflection of sandwich plates with orthotropic cores, *AIAA J.* **2**, 10, Oct. 1964, pp. 1820–2.

Struts; Wrinkling

6.1. K. H. BOLLER and C. B. NORRIS, *Elastic Stability of the Facings of Flat Sandwich Panels when Subjected to Edgewise Stresses*, FPL Report 1802, Feb. 1949.
6.2. B. G. HEEBINK and A. A. MOHAUPT, *Effect of Defects of Strength on Aircraft-type Sandwich Panels*, FPL Report 1809, Sept. 1949. Also FPL Report D1809A, Nov. 1951.
6.3. C. B. NORRIS et al., *Wrinkling of the Facings of Sandwich Constructions Subjected to Edgewise Compression*, FPL Report 1810, Nov. 1949.
6.4. C. B. NORRIS, K. H. BOLLER and A. W. VOSS, *Wrinkling of the Facings of Sandwich Construction Subjected to Edgewise Compression. Sandwich constructions having honeycomb cores*, FPL Report 1810-A, June 1953.
6.5. K. H. BOLLER and C. B. NORRIS, *Effect of Shear Strength on Maximum Loads of Sandwich Columns*, FPL Report 1815, June 1950.
6.6. C. B. NORRIS and W. J. KOMMERS, *Short-column Compressive Strength of Sandwich Constructions as affected by the Size of the Cells of Honeycomb-core Materials*, FPL Report 1817, Aug. 1950.
6.7. C. B. NORRIS, *Effect of Unbonded Joints in an Aluminium Honeycomb-core Material for Sandwich Construction*, FPL Report 1835, May 1952.
8.1. D. WILLIAMS, D. M. A. LEGGETT and H. G. HOPKINS, *Flat Sandwich panels under Compressive End Loads*, A.R.C., R & M 1987, 1941.
8.2. H. L. COX and J. R. RIDDELL, *Sandwich Construction and Core Materials*, Part III. Instability of sandwich struts and beams, A.R.C., R & M 2125, 1945.
8.3. F. T. BARWELL and J. R. RIDDELL, *The Wrinkling of Sandwich Struts*, A.R.C., R & M 2143, 1946.
9.1. N. J. HOFF and S. E. MAUTNER, Buckling of sandwich-type panels, *J. Aero. Sci.* **12**, 3, July 1945, pp. 285–97.
9.2. C. C. WAN, Face buckling and core strength requirements in sandwich construction, *J. Aero. Sci.* **14**, 9, Sept. 1947, pp. 531–9.
9.3. J. N. GOODIER and I. M. NEOU, The evaluation of theoretical critical compression in sandwich plates, *J. Aero. Sci.* **18**, 10, Oct. 1951, pp. 649–57.
9.4. J. N. GOODIER and C. S. HSU, Nonsinusoidal buckling modes of sandwich plates, *J. Aero. Sci.* **21**, 8, Aug. 1954, pp. 525–32.
10.1. G. S. GOUGH, C. F. ELAM and N. D. DE BRUYNE, The stabilization

of a thin sheet by a continuous supporting medium, *J. Roy. Aero. Soc.* **44**, 349, Jan. 1940, pp. 12–43.

10.2. J. N. GOODIER, Cylindrical buckling of sandwich plates, *J. App. Mech.* **13**, 4, Dec. 1946, pp. A253–60.

10.3. H. W. MARCH, Elastic stability of the facings of sandwich columns, *Proc. Symp. App. Maths.* **3** *Elasticity*, pp. 85–106, McGraw Hill, 1950.

10.4. S. YUSSUFF, Theory of wrinking in sandwich construction, *J. Roy. Aero. Soc.* **59**, 529, Jan. 1955, pp. 30–36.

10.5. S. YUSSUFF, Face wrinkling and core strength in sandwich construction, *J. Roy. Aero. Soc.* **64**, 591, Mar. 1960, pp. 164–7.

Core Materials: Properties and Methods of Testing

11.1. ANON., *Tentative Methods of Test for Determining Strength Properties of Core Material for Sandwich Construction*, FPL Report 1555 Dec. 1946, revised Oct. 1948.

11.2. ANON., *Methods for Conducting Mechanical Tests of Sandwich Constructions at Normal Temperatures*, FPL Report 1556, revised Feb. 1950.

11.3. W. J. KOMMERS, *Strength Properties of Plastic Honeycomb Core Materials*, FPL Report 1805, Dec. 1949.

11.4. A. W. VOSS, Mechanical properties of some low-density materials and sandwich cores, FPL Report 1826, Mar. 1952.

11.5. E. W. KUENZI, *Mechanical Properties of Aluminium Honeycomb Cores*, FPL Report 1849, Sept. 1955.

11.6. E. W. KUENZI and V. C. SETTERHOLM, *Mechanical Properties of Aluminium Multiwave Cores*, FPL Report 1855, Sept. 1956.

11.7. E. W. KUENZI, *Mechanical Properties of Glass-fabric Honeycomb Cores*, FPL Report 1861, Mar. 1957.

11.8. E. W. KUENZI and W. E. JAHNKE, *Mechanical Properties of Some Heat-resistant Metal Honeycomb Cores*, FPL Report 1872, Dec. 1959.

11.9. G. H. STEVENS and E. W. KUENZI, *Mechanical Properties of Several Honeycomb Cores*, FPL Report 1887, July 1962.

11.10. K. H. BOLLER, *Durability of Resin-treated Paper Honeycomb Core*, FPL Report 2158, Nov. 1959.

12.1. C. B. NORRIS, *An Analysis of the Compressive Strength of Honeycomb Cores for Sandwich Construction*, NACA TN 1251, 1947.

12.2. F. WARREN and C. B. NORRIS, *Analysis of Shear Strength of Honeycomb Cores for Sandwich Construction*, NACA TN 2208, 1950.

12.3. L. A. RINGELSTETTER, A. W. VOSS and C. B. NORRIS, *Effect of Cell Shape on Compressive Strength of Hexagonal Honeycomb Structures*, NACA TN 2243, 1950.

12.4. C. LIBOVE and R. E. HUBKA, *Elastic Constants for Corrugated Core Sandwich Plates*, NACA TN 2289, 1951.

12.5. M. S. ANDERSON, *Local Instability of the Elements of a Truss-core*

Sandwich Plate, NACA TN 4292, 1958. Also NASA Tech. Report R.30, 1959.

13.1. F. T. BARWELL, *Sandwich Construction and Core Materials*. Part I. *An Introduction to sandwich construction*, A.R.C., R & M 2123, Dec. 1945.

13.2. N. E. TOPP, *Sandwich Construction and Core Materials*. Part II. *The preparation of low-density core materials for use as cores in sandwich construction*, A.R.C., R & M 2124, Dec. 1945.

13.3. F. T. BARWELL, *Sandwich Construction and Core Materials*. Part IV. *Notes on some methods of testing core materials*. A.R.C., R & M 2467, 1946.

13.4. W. J. PULLEN, R. G. CHAPMAN, S. PEARSON and J. K. OAKS, *Sandwich Construction and Core Materials*. Part VI., A.R.C., R & M 2687, 1948.

14.1. G. A. HOFFMAN, Poisson's ratio for honeycomb sandwich cores, *J. Aero. Sci.* **25**, 8, Aug. 1958, pp. 534–5.

15.1. ANON., *Compressive Strength, Edgewise, of Flat Sandwich Constructions*, ASTM Standard C364.

15.2. ANON., *Compressive Strength, Flatwise, of Sandwich Cores*, ASTM Standard C365.

15.3. ANON., *Flexure Test of Flat Sandwich Constructions*, ASTM Standard C393.

15.4. ANON., *Climbing Drum Peel Test for Adhesives*, ASTM Standard D1781.

15.5. ANON., *Shear Test in Flatwise Plane of Flat Sandwich Construction or Sandwich Cores*, ASTM Standard C273.

15.6. ANON., *Tension Test of Flat Sandwich Constructions in Flatwise Plane*. ASTM Standard C297.

15.7. ANON., *Compressive Properties of Rigid Plastics*, ASTM Standard D695.

15.8. ANON., *Testing Veneer, Plywood and other Glued Veneer Constructions*, ASTM Standard D805.

15.9. W. G. PLUMTREE and W. CHEORVAS, Effect of dimensional factors and temperature on the shear strength of aluminium honeycomb, *Symp. Structural Sandwich Constructions*. ASTM STP 201, 1956, pp. 13–22.

15.10. S. KELSEY, R. A. GELLATLY and B. W. CLARK, The shear modulus of foil honeycomb core, *Aircraft Engg.* **30**, 356, Oct. 1958, pp. 294–302.

15.11. A. D. SAPOWITH, Transverse shear stiffness for the double 'V' corrugated-core sandwich panel, *Aerospace Engg.* **18**, 9, Sept. 1959, pp. 53–56.

15.12. W. G. YOUNGQUIST and E. W. KUENZI, Shear and torsion testing of wood, plywood and sandwich constructions at the U.S. Forest Products Laboratory, *Symp. Shear and Torsion Testing*, ASTM STP 289, 1960, pp. 75–89.

15.13. H. P. O'SULLIVAN, Double block shear test for foil honeycomb cores, *Aircraft Engg.* **33**, 385, Mar. 1961, pp. 64–66.

15.14. C. C. CHANG and I. K. EBCIOGLU, Effect of cell geometry on the shear modulus and on density of sandwich panel cores, *Trans. ASME* **83** (2), Series D, Dec. 1961, pp. 513–18.

15.15. ———, *Expanded Plastics* (translation from Russian), Pergamon, 1963.

15.16. T. H. FERRIGNO, *Rigid Plastics Foams*, Reinhold, 1963.

15.17. J. PENZIEN and T. DIDRIKSSON, Effective shear modulus of honeycomb cellular structure, *AIAA J.* **2**, 3, Mar. 1964, pp. 531–5.

15.18. B. HUGHES and R. L. WAJDA, Plastics sandwich panels with various foamed core materials, and their behaviour under load, *Symp. Plastics in Building Structures*, Pergamon, London, 1965, pp. 209–20.

Sandwich Beams; Methods of Testing Beams
See also ref. 15.3.

16.1. H. W. MARCH and C. B. SMITH, *Flexural Rigidity of a Rectangular Strip of Sandwich Construction*, FPL Report 1505, Feb. 1944, revised Feb. 1955.

16.2. C. B. NORRIS, W. S. ERICKSEN and W. J. KOMMERS, *Supplement to Flexural Rigidity of a Rectangular Strip of Sandwich Construction*, FPL Report 1505A, 1944, revised May 1952.

16.3. E. W. KUENZI, *Edgewise Compression Strength of Panels and Flatwise Flexural Strength of Strips of Sandwich Construction*, FPL Report 1827, Nov. 1951.

16.4. E. W. KUENZI, *Flexure of Structural Sandwich Construction*, FPL Report 1829, Dec. 1951.

19.1. N. J. HOFF and S. E. MAUTNER, Bending and buckling of sandwich beams, *J. Aero. Sci.* **15**, 12, Dec. 1948, pp. 707–20.

20.1. H. B. HOWARD, The five-point loading shear stiffness test, *J. Roy. Aero. Soc.* **66**, 621, Sept. 1962, p. 591.

20.2. H. L. COX and D. W. MARTIN, Deformation of sandwich structure, *J. Roy. Aero. Soc.* **69**, 651, Mar. 1965, pp. 193–7.

20.3. H. G. ALLEN, Measurement of shear stiffness of sandwich beams, *Trans. J. Plastics Inst.* **35**, 115, Feb. 1967, pp. 359–63.

Optimum Design

21.1. E. W. KUENZI, *Minimum weight structural sandwich*, U.S. Forest Service Res. Note FPL-086, Jan. 1965.

22.1. R. E. HUBKA, N. F. DOW and P. SEIDE, *Relative Structural Efficiences of Flat Balsa-core Sandwich and Stiffened-panel Construction*, NACA TN 2514, 1951.

22.2. A. E. JOHNSON and J. W. SEMONIAN, *A Study of the Efficiency of High-strength Steel, Cellular-core Sandwich Plates in Compression*, NACA TN 3751, 1956.

22.3. M. S. ANDERSON, *Optimum Proportions of Truss-core and Web-core Sandwich Plates Loaded in Compression*, NASA TN D-98, 1959.
23.1. P. ACKERS, *Efficiency of Sandwich Struts using a Calcium Alginate Core*, A.R.C., R & M 2015, 1945.
24.1. W. H. WITTRICK, *A Theoretical Analysis of the Efficiency of Sandwich Construction under Compressive End Load*, A.R.C., R & M 2016, 1945.
25.1. P. P. BIJLAARD, On the optimum distribution of material in sandwich plates loaded in their plane, *Proc. 1st U.S. Nat. Congr. App. Mech.*, 1951, pp. 373-80.
25.2. W. FLUEGGE, The optimum problem of the sandwich plate, *J. App. Mech.* **19**, 1, Mar. 1952, pp. 104-8.
25.3. H. G. ALLEN, Optimum design of sandwich struts and beams, *Symp. Plastics in Building Structures*, London, 1965. Pergamon.

Reviews, Surveys, Bibliographies
See also ref. 5.12.

26.1. ANON., *List of Publications on Mechanical Properties and Structural Uses of Wood and Wood Products*, [Includes sandwich construction], FPL No. 200, Dec. 1962.
27.1. M. S. ANDERSON and R. G. UPDEGRAFF, *Some Research Results on Sandwich Structures*, NACA TN 4009, 1957.
28.1. D. WILLIAMS, *Sandwich Construction. A Practical Approach for the Use of Designers*, A.R.C., R & M 2466, 1947.
30.1. A. GARRARD, Theory of sandwich construction, *British Plastics* **18**, 208 and 209, Sept. and Oct. 1946, pp. 380-8 and 451-8.
30.2. ———, *Symposium on Structural Sandwich Constructions*, [reviews by various authors], A.S.T.M., S.T.P. 118, 1951.
30.3. J. SOLVEY, *Bibliography and Summaries of Sandwich Constructions (1939-54)*, Aero. Res. Lab. (Australia), ARL/SM2, Oct. 1955.
30.4. ANON., *Sandwich Construction*, OTS Selective Bibliography SB-517, Sept. 1963, U.S. Dept. of Commerce.
30.5. L. M. HABIP, A review of recent Russian work on sandwich construction, *Int. J. Mech. Sci.* **6**, 6, Dec. 1964, pp. 483-7.
30.6. L. M. HABIP, A survey of modern developments in the analysis of sandwich structures, *App. Mech. Rev.* **18**, 2, Feb. 1965, pp. 93-98.
30.7. ANON., *Honeycomb and Sandwich Materials*, Review OTR-119, July 1965, U.S. Dept. of Commerce.

Unclassified References

31.1. L. W. WOOD, *Sandwich Panels for Building Construction*, FPL Report 2121, Oct. 1958.
31.2. L. J. MARKWARDT and L. W. WOOD, *Long-term Case Study of Sandwich Panel Construction in FPL Experimental Unit*, FPL Report 2165, Oct. 1959.

REFERENCES

35.1. W. Fairbairn, *An Account of the Construction of the Britannia and Conway Tubular Bridges*, John Weale et al., London, 1849.

35.2. L. N. G. Filon, On antiplane stress in an elastic solid, *Proc. Roy. Soc.* A, **160,** 1937, pp. 137–54.

35.3. S. Timoshenko, *Theory of Plates and Shells*, McGraw-Hill, 1940.

35.4. J. Lockwood Taylor, Notes on sandwich construction, *Aircraft Engg.* **21,** 244, June 1949, p. 196.

35.5. ———, *Symposium on Structural Sandwich Constructions*, ASTM STP 118, 1951.

35.6. J. N. Goodier, Some observations on elastic stability, *Proc. 1st U.S. Nat. Congr. App. Mech. 1951*, pp. 193–202.

35.7. G. Horvey, Bending of honeycombs and of perforated plates, *J. App. Mech.* **19,** 1, Mar. 1952, pp. 122–3.

35.8. P. Seide, On the torsion of rectangular sandwich plates, *J. App. Mech.* **23,** 2, June 1956, pp. 191–4.

35.9. ———, *Symposium on Structural Sandwich Constructions* [Papers related to constructional details], ASTM STP 201, 1956.

35.10. ———, *Symposium on Durability and Weathering of Structural Sandwich Constructions*, ASTM STP 270, 1959.

35.11. Yi-Yuan Yu, A new theory of elastic sandwich plates, *J. App. Mech.* **26,** E3, Sept. 1959, pp. 415–21.

35.12. Anon., *Marine Design Manual for Fiberglass Reinforced Plastics*, McGraw-Hill, N.Y., 1960.

35.13. Shun Cheng, A formula for the torsional stiffness of rectangular sandwich plates, *J. App. Mech.* **28,** 2, June 1961, pp. 307–9.

35.14. S. Timoshenko and J. M. Gere, *Theory of Elastic Stability*, McGraw-Hill, 2nd ed., 1961.

35.15. G. Gerard, *Introduction to Structural Stability Theory*, McGraw-Hill, 1962.

35.16. I. K. Ebcioglu, On a theory of sandwich panels in the reference state, *Int. J. Engg. Sci.* **2,** 6, Mar. 1965, pp. 549–64.

INDEX

Anticlastic bending 47, 266
Antiplane core xv, 14

Beam-columns with thick faces
  and non-sinusoidal transverse load 71, 72
  and sinusoidal transverse load 65–71
  and uniform transverse load 72–75
Beams
  assumptions 8–14, 217–20
  formulae summarized 220, 221
  long or short 32, 219
  narrow or wide 46, 220, 267
  optimum design of 237, 238, 241–4
  sign convention 9, 15
  with thick faces
    and core stiff in bending 43–46
    and distributed load 33–36, 221
    and four-point load 36–41
    and point load 25–33, 221
    and unequal faces 41–43, 232–3
    and unsymmetric load 24, 25
  with thin faces
    and distributed load 18, 221, 237
    and point load 17, 18, 220
    and unequal faces 220, 232
    and unsymmetric load 19–21
Boundary conditions
  for beams with thick faces 26–28, 34, 38, 207
  for plates 97, 98, 207–8
  for struts 50, 51, 54, 56

Columns see Struts
Convergence of series 74, 94
Core materials
  properties of 138, 248–54
  tests 254–7
  types 1–7
Corrugated cores, properties of 5, 96, 138, 253
Cylindrical bending 46, 266

Design
  of core thickness 236–41
  of core and face thickness 236, 241–4
  of face thickness and core thickness and density 236
Differential equations
  for beam with thick faces 24
  for beam-column with thick faces 269
  for plates 132, 134, 135, 191
  for struts 49, 51
Dimpling see Intracellular buckling

Face bending stiffness in relation to core shear stiffness 21, 219
Face materials 3, 245
Face thickness
  definitions xv, xvi, 219

INDEX

Face thickness (*cont.*)
 effect in beams 10, 32, 217, 221
 plates 118, 119, 124, 135, 136, 204–6
 struts 52, 53, 222
 unequal 41, 137, 232, 233
Five-point load test 260
Four-point load test 38, 39, 246, 247, 261

Honeycomb cores 3–5, 248–52, 254

Interaction formulae 50, 52, 90, 198
Intracellular buckling 180, 225

Lagrangian multipliers 210
Levy's method 210
Local instability 179, 180, 225, 226

Minimum weight or cost *see* Design

Narrow beams *see* Beams
Notations
 for beams and plates xii–xv, 15, 78, 194, 200–7
 for honeycomb or wood cores 250

Optimum design *see* Design

Parameters defining character of sandwich xv, 24, 30, 55, 66, 67, 87, 91, 117, 119, 135, 136, 162, 166, 176, 188, 213, 223, 226, 233
Plates
 bending and shear displacements compared 91
 bending results 91–94, 120–4, 150–5, 229–31
 buckling results (P_x) 88, 89, 118, 119, 143, 144, 146–50, 226–9
 combined transverse and edgewise loads 94, 95, 147–51, 232, 233
 design of *see* Design
 edgewise loads in general 211, 212, 233
 equality of λ and μ (isotropic case) 87
 formulae for design parameters xv, 87, 117, 119, 137, 226, 233
 homogeneous, comparison with 94
 initially deformed 213
 large deflections 212, 213
 not simply supported 207–11, 233
 sign convention 78
 very weak cores 124, 125, 227, 228
Poisson's ratios
 measurement of 262
 of orthotropic solids 264
 of sandwich plates 129
Potential energy expressions
 for end load on strut 64
 for loads on plates 83, 107
 for transverse loads on beams 65

Rayleigh–Ritz procedure 66, 86, 109–11, 209

Series, convergence of 74, 94
Sign conventions 15, 78
Stiffness
 bending
  of beams xv, 9, 13, 22, 217, 232, 241
  of plates xv, 129, 137, 233, 243

Stiffness (*cont.*)
  shear
    of beams xv, 17, 24, 30, 43–46, 218, 219
    of plates 130, 135, 137, 138
    torsional, of plates 129, 137, 138, 262
Strain energy expressions
  for beams and struts 61–63
  for orthotropic solids 266
  for plates 80–82, 104–6, 201
  of Hoff 201
  of Libove and Batdorf 201
Struts
  boundary conditions for 50, 51, 54, 56
  buckling curves for 53, 186, 187
  critical loads of 48, 50, 51–53, 184, 222
  design of 238, 241
  formulae for, summarized 222
  narrow and wide 221
  shear buckling load 50, 52, 173
  unequal faces 53

Tests
  climbing drum (peel) 255
  for core materials
    in shear 255–7
    in transverse tension and compression 254
  for face materials 246
  for sandwich beams 36–41, 246, 247, 257–60
  for sandwich plates (torsion) 262
Three-point load test 25–33, 247, 257–60

Wide beams *see* Beams
Wrinkling instability
  formulae summarized 223
  interaction with overall instability 181–9
  non-sinusoidal 200
  stress functions for core 164, 165
  with antiplane core 171–3
  with faces in biaxial stress 179
  with initially deformed faces 173–8, 224
  with isotropic core
    core on rigid base 160–3, 224
    core semi-infinite 157–60
    symmetric and antisymmetric 163–7, 223, 224
  with orthotropic core 178, 225
  with Winkler core (springs) 169–71

Printed in Germany
by Amazon Distribution
GmbH, Leipzig